DEEP
MEDICINE

DEEP MEDICINE

How Artificial Intelligence
Can Make Healthcare Human Again

ERIC TOPOL

BASIC BOOKS

New York

Basic Books
Hachette Book Group
1290 Avenue of the Americas, New York, NY 10104
www.basicbooks.com

Printed in the United States of America
First Edition: March 2019
Published by Basic Books, an imprint of Perseus Books, LLC, a subsidiary of Hachette Book Group, Inc. The Basic Books name and logo is a trademark of the Hachette Book Group.

The Hachette Speakers Bureau provides a wide range of authors for speaking events. To find out more, go to www.hachettespeakersbureau.com or call (866) 376-6591.

The publisher is not responsible for websites (or their content) that are not owned by the publisher.

Print book interior design by Linda Mark.

The Library of Congress has cataloged the hardcover edition as follows:
  Names: Topol, Eric J., 1954– author.
Title: Deep medicine : how artificial intelligence can make healthcare human again / Eric Topol.
Description: First edition. | New York : Basic Books, March 2019. | Includes bibliographical references and index.
Identifiers: LCCN 2018043186 (print) | LCCN 2018043932 (ebook) | ISBN 9781541644649 (ebook) | ISBN 9781541644632 (hardcover)
Subjects: | MESH: Artificial Intelligence | Medical Informatics | Diagnosis, Computer-Assisted | Therapy, Computer-Assisted | Quality Improvement
Classification: LCC R858 (ebook) | LCC R858 (print) | NLM W 26.55.A7 | DDC 362.10285—dc23
LC record available at https://lccn.loc.gov/2018043186

ISBNs: 978-1-5416-4463-2 (hardcover); 978-1-5416-4464-9 (ebook)

LSC-C

10 9 8 7 6 5 4 3 2 1

*To my family—Susan, Sarah, Evan, Antonio,
Julian, and Isabella—who provided unconditional support
and the deep inspiration for me to pursue this work*

CONTENTS

FOREWORD

Life can only be understood backwards;
but it must be lived forwards.

—SØREN KIERKEGAARD

AMONG THE MANY CHARACTERISTICS THAT MAKE US HUMAN
and that distinguish us from other animals must be our urge
to look back. It is hard to imagine that other species brood late at
night about the one that got away or a job they could have had. But
we also do it as a form of scholarship, looking back at ourselves as a
species, as if we were the Creator, poring through recorded history,
charting the milestones of progress, from the harnessing of fire to the
microchip. Then we try to make sense of it.

Kierkegaard's thesis that we live life forward but understand it
backward might mean nothing more than we remember the past,
and at best we have an (inaccurate) record of it. But with apologies
to him and to George Santayana, understanding history does not

provide immunity to repeating it. A cursory scan of the news shows this to be true. In short, even as a guide to what to avoid, the past is unreliable. Only the future is certain because it is still ours to make.

Which brings us to futurists, like the author of this wonderful book. Such individuals, on hearing that the Wright brothers became airborne, can foresee budget airlines, airline hubs, and humans walking on the moon. These historians of the now begin with the study of what *is* today, asking not how to avoid the perils of the past but how to maximize the advantages of the future. Pencil and paper, or tablet, in hand, they patrol the frontiers of science and tech and interview those at the cutting edge, including those who have tumbled over. They seek out innovators, scientists, mavericks, and dreamers. They listen, they monitor, they filter, and they synthesize knowledge across many disciplines to make sense of it all for the rest of us. As *Deep Medicine* will show you, theirs is a formidable intellectual task and an extraordinarily creative one. It involves as much right brain as left, and it invokes the muses, because what is in this book is as much inspiration as it is exposition.

Deep Medicine is Eric Topol's third exploration of what *will* be. The previous books, examined in the light of where we are now, reveal his prescient vision. In *Deep Medicine,* Eric tells us we are living in the Fourth Industrial Age, a revolution so profound that it may not be enough to compare it to the invention of steam power, the railroads, electricity, mass production, or even the computer age in the magnitude of change it will bring. This Fourth Industrial Age, revolving around artificial intelligence (AI), robotics, and Big Data, heralds a profound revolution that is already visible in the way we live and work, perhaps even in the way we think of ourselves as humans. It has great potential to help, but also to harm, to exaggerate the profound gap that already exists between those who have much and those who have less each passing year.

This revolution will overtake every human endeavor, medicine not least among them. Medicine itself is at a moment of crisis. As a profession, for all the extraordinary advances in the art and science of medicine in the last four decades, we have too often failed our patients. We fail to follow proven guidelines, and we fail in the art by not seeing the unique *person* in front of us. We know their genome, but by not *listening* to their story, we don't register their broken heart. We fail to see the neurofibroma that are raising lumps all over their skin, a finding that is relevant to their paroxysmal hypertension but that does need the gown to come off during the exam, does need our attention to be on the body and not on the screen; we miss the incarcerated hernia that explains an elderly patient's vomiting and have to wait for an expensive CAT scan and a radiologist to tell us what was before our eyes. Countries with the biggest expenditures on healthcare lag behind those that spend much less in basic rankings such as infant mortality. I think it is very telling that *Deep Medicine* opens with a profound, personal, revealing anecdote of the author's own painful and harrowing medical encounter that was a result of not being seen as an individual, someone with an uncommon disorder.

It should not surprise us that technology, despite the dramatic way it has altered our ability to image the body, to measure and monitor its molecular structure, can also fail just as badly as humans fail. The glaring example is in the electronic healthcare record systems (EHRs) currently in use in most hospitals. These EHRs were designed for billing, not for ease of use by physicians and nurses. They have affected physician well-being and are responsible for burnout and attrition; moreover, they have forced an inattentiveness to the patient by virtue of an intruder in the room: the screen that detracts from the person before us. In *Intoxicated by My Illness,* a poignant memoir about a man's ultimately fatal prostate cancer, Anatole Broyard articulates a wish that his urologist would "brood on my situation for perhaps five minutes, that he would give me his whole mind just once, be bonded with me for a brief space, survey my soul as well as my flesh, to get at my illness, *for each man is ill in his own way.*"[1] This poignant declaration, from the era just before electronic medical records, expresses the fundamental need of a sick human being; it is timeless, I believe, resistant to change, even as the world around us changes. It bears emphasizing: *each man and woman is ill in his or her own way.*

I am excited about the future, about the power to harness Big Data. By their sheer capacity to plow through huge datasets and to learn as they go along, artificial intelligence and deep learning will bring tremendous precision to diagnosis and prognostication. This isn't to say they will replace humans: what those technologies will provide is a recommendation, one that is perhaps more accurate than it has ever been, but it will take a savvy, caring, and attentive physician and healthcare team to tailor that recommendation to— and with—the individual seated before them. Over 2,000 years ago, Hippocrates said, "It is more important to know what sort of person has [a] disease than to know what sort of disease a person has." In a 1981 editorial on using a computer to interpret risk after exercise stress testing, Robert Califf and Robert Rosati wrote, "Proper interpretation and use of computerized data will depend as much on wise doctors as any other source of data in the past."[2] This is a timeless principle, so long as it is humans we are discussing and not brake parts on an assembly line.

We come back in the end to the glorious fact that we are human, that we are embodied beings, a mind with all its complexities in a body that is equally complex. The interplay between one and the other remains deeply mysterious. What is not mysterious is this: when we are ill, we have a fundamental need to be *cared* for; disease infantilizes us, particularly when it is severe, and though we want the most advanced technical skills, scientific precision, the best therapy, and though we would want our physicians to "know" us (and unlike the time of Hippocrates, such knowing includes the genome, proteome, metabolome, transcriptome, predictions driven by AI, and so on), we badly want it to be expressed in the form of a caring, conscientious physician and healthcare team. We want the physician—a caring individual and not a machine—to give us time, to perform an attentive exam if for no other reason than to acknowledge the locus of disease on our body and not on a biopsy or an image or a report, to validate our personhood and our complaint by touching where it hurts. As Peabody said years ago, the secret of caring for patients is in caring for the patient.

We want those who care for us to know our hearts, our deepest fears, what we live for and would die for.

That is, and it always will be, our deepest desire.

Abraham Verghese, MD
Department of Medicine
Stanford University

chapter one

INTRODUCTION TO DEEP MEDICINE

> By these means we may hope to achieve not indeed a brave
> new world, no sort of perfectionist Utopia, but the more
> modest and much more desirable objective—a genuinely
> human society.
>
> —ALDOUS HUXLEY, 1948

"YOU SHOULD HAVE YOUR INTERNIST PRESCRIBE ANTI-depression medications," my orthopedist told me.

My wife and I looked at each other, bug-eyed, in total disbelief. After all, I hadn't gone to my one-month post-op clinic visit following a total knee replacement seeking psychiatric advice.

My knees went bad when I was a teenager because of a rare condition known as osteochondritis dissecans. The cause of this disease remains unknown, but its effects are clear. By the time I was twenty years old and heading to medical school, I had already had dead bone sawed off and extensive reparative surgery in both knees. Over the next forty years, I had to progressively curtail my physical activities, eliminating running, tennis, hiking, and elliptical exercise. Even walking became painful, despite injections of steroids and synovial fluid directly into the knee. And so at age sixty-two I had my left knee replaced, one of the more than 800,000 Americans who have this surgery, the most common orthopedic operation. My orthopedist had deemed me a perfect candidate: I was fairly young, thin, and fit. He said the only significant downside was a 1 to 2 percent risk of infection. I was about to discover another.

After surgery I underwent the standard—and, as far as I was told, only—physical therapy protocol, which began the second day after surgery. The protocol is intense, calling for aggressive bending and extension to avoid scar formation in the joint. Unable to get meaningful flexion, I put a stationary bicycle seat up high and had to scream in agony to get through the first few pedal revolutions. The pain was well beyond the reach of oxycodone. A month later, the knee was purple, very swollen, profoundly stiff, and unbending. It hurt so bad that I couldn't sleep more than an hour at a time, and I had frequent crying spells. Those were why my orthopedist recommended antidepressants. That seemed crazy enough. But the surgeon then recommended a more intensive protocol of physical therapy, despite the fact that each session was making me worse. I could barely walk out of the facility or get in my car to drive home. The horrible pain, swelling, and stiffness were unremitting. I became desperate for relief, trying everything from acupuncture, electro-acupuncture, cold laser, an electrical stimulation (TENS) device, topical ointments, and dietary supplements including curcumin, tart cherry, and many others—fully cognizant that none of these putative treatments have any published data to support their use.

Joining me in my search, at two months post-op, my wife discovered a book titled *Arthrofibrosis*. I had never heard the term, but it turned out to be what I was suffering from. Arthrofibrosis is a complication that occurs in 2 to 3 percent of patients after a knee replacement—that makes the condition uncommon, but still more common than the risk of infection that my orthopedist had warned me about. The first page of the book seemed to describe my situation perfectly: "Arthrofibrosis is a disaster," it said. More specifically, arthrofibrosis is a vicious inflammation response to knee replacement, like a rejection of the artificial joint, that results in profound scarring. At my two-month post-op visit, I asked my orthopedist whether I had arthrofibrosis. He said absolutely, but there was little he could do for the first year following surgery—it was necessary to allow the inflammation to "burn out" before he could go back in and remove the scar tissue. The thought of going a year as I was or having another operation was making me feel even sicker.

Following a recommendation from a friend, I went to see a different physical therapist. Over the course of forty years, she had seen many patients with osteochondritis dissecans, and she knew that, for patients such as me, the routine therapeutic protocol was the worst thing possible. Where the standard protocol called for extensive, forced manipulation to maximize the knee flexion and extension (which was paradoxically stimulating more scar for-mation), her approach was to go gently: she had me stop all the weights and exercises and use anti-inflammatory medications. She handwrote a page of instructions and texted me every other day to ask how "our knee" was doing. Rescued, I was quickly on the road to recovery. Now, years later, I still have to wrap my knee every day to deal with its poor healing. So much of this torment could have been prevented.

As we'll see in this book, artificial intelligence (AI) could have predicted that my experience after the surgery would be complicated. A full literature review, provided that experienced physical thera-pists such as the woman I eventually found shared their data, might well have indicated that I needed a special, bespoke PT protocol. It wouldn't only be physicians who would get a better awareness of the risks confronting their patients. A virtual medical assistant, residing in my smartphone or my bedroom, could warn me, the patient, directly of the high risk of arthrofibrosis that a standard course of physical therapy posed. And it could even tell me where I could go to get gentle rehab and avoid this dreadful problem. As it was, I was blindsided, and my orthopedist hadn't even taken my history of osteochondritis dissecans into account when discussing the risk of surgery, even though he later acknowledged that it had, in fact, played a pivotal role in the serious problems that I encountered.

Much of what's wrong with healthcare won't be fixed by advanced technology, algorithms, or machines. The robotic response of my doctor to my distress exemplifies the deficient component of care. Sure, the operation was done expertly, but that's only the technical component. The idea that I should take medication for depression exemplifies a profound lack of human connection and empathy in medicine today. Of course, I was emotionally depressed, but depression wasn't the problem at all: the problem was that I was in severe pain and had Tin Man immobility. The orthopedist's lack of compassion was palpable: in all the months after the surgery, he never contacted me once to see how I was getting along. The physical therapist not only had the medical knowledge and experience to match my condition, but she really cared about me. It's no wonder that we have an opioid epidemic when it's a lot quicker and easier for doctors to prescribe narcotics than to listen to and understand patients.

Almost anyone with chronic medical conditions has been "roughed up" like I was—it happens all too frequently. I'm fortunate to be inside the medical system, but, as you have seen, the problem is so pervasive that even insider knowledge isn't necessarily enough to guarantee good care. Artificial intelligence alone is not going to solve this problem on its own. We need humans to kick in. As machines get smarter and take on suitable tasks, humans might actually find it easier to be more humane.

AI in medicine isn't just a futuristic premise. The power of AI is already being harnessed to help save lives. My close friend, Dr. Stephen Kingsmore, is a medical geneticist who heads up a pioneering program at the Rady Children's Hospital in San Diego. Recently, he and his team were awarded a Guinness World Record for taking a sample of blood to a fully sequenced and interpreted genome in only 19.5 hours.[1]

A little while back, a healthy newborn boy, breastfeeding well, went home on his third day of life. But, on his eighth day, his mother brought him to Rady's emergency room. He was having constant seizures, known as status epilepticus. There was no sign of infection. A CT scan of his brain was normal; an electroencephalogram just showed the electrical signature of unending seizures. Numerous potent drugs failed to reduce the seizures; in fact, they were getting even more pronounced. The infant's prognosis, including both brain damage and death, was bleak.

A blood sample was sent to Rady's Genomic Institute for a rapid whole-genome sequencing. The sequence encompassed 125 gigabytes of data, including nearly 5 million locations where the child's genome differed from the most common one. It took twenty seconds for a form of AI called natural-language processing to ingest the boy's electronic medical record and determine eighty-eight phenotype features (almost twenty times more than the doctors had summarized in their problem list). Machine-learning algorithms quickly sifted the approximately 5 million genetic variants to find the roughly 700,000 rare ones. Of those, 962 are known to cause diseases. Combining that information with the boy's phenotypic data, the system identified one, in a gene called ALDH7A1, as the most likely culprit. The variant is very rare, occurring in less than 0.01 percent of the population, and causes a metabolic defect that leads to seizures. Fortunately, its effects can be overridden by dietary supplementation with vitamin B6 and arginine, an amino acid, along with restricting lysine, a second amino acid. With those changes to his diet made, the boy's seizures abruptly ended, and he was discharged home thirty-six hours later! In follow-up, he is perfectly healthy with no sign of brain damage or developmental delay.

The key to saving this boy's life was determining the root cause of his condition. Few hospitals in the world today are sequencing the genomes of sick newborns and employing artificial intelligence to make everything known about the patient and genomics work together. Although very experienced physicians might eventually have hit upon the right course of treatment, machines can do this kind of work far quicker and better than people.

So, even now, the combined efforts and talents of humans and AI, working synergistically, can yield a medical triumph. Before we get too sanguine about AI's potential, however, let's turn to a recent experience with one of my patients.

"I want to have the procedure," my patient told me on a call after a recent visit.

A white-haired, blue-eyed septuagenarian who had run multiple companies, he was suffering from a rare and severe lung condition known as idiopathic—a fancy medical word for "of unknown cause"— pulmonary fibrosis. It was bad enough that he and his pulmonologist had been considering a possible lung transplant if it got any worse. Against this backdrop he began to suffer a new symptom: early-onset fatigue that left him unable to walk more than a block or swim a lap. He had seen his lung doctor and had undergone pulmonary function tests, which were unchanged. That strongly suggested his lungs weren't the culprit.

He, along with his wife, then came to see me, very worried and depressed. He took labored, short steps into the exam room. I was struck by his paleness and look of hopelessness. His wife corroborated his description of his symptoms: there had been a marked diminution of his ability to get around, to even do his daily activities, let alone to exert himself.

After reviewing his history and exam, I raised the possibility that he might have heart disease. A few years previously, after he began to suffer calf pain while walking, he had stenting of a blockage in his iliac artery to the left leg. This earlier condition raised my concern about a cholesterol buildup in a coronary artery, even though he had no risk factors for heart disease besides his age and sex, so I ordered a CT scan with dye to map out his arteries. The right coronary artery showed an 80 percent narrowing, but the other two arteries were free of significant disease. It didn't fit together. The right coronary artery doesn't supply very much of the heart muscle, and, in my thirty years as a cardiologist (twenty of which involved opening coronary arteries), I couldn't think of any patients with such severe fatigue who had narrowing in only the right coronary artery.

I explained to him and to his wife that I really couldn't connect the dots, and that it might be the case of a "true-true, unrelated"— that the artery's condition might have nothing to do with the fatigue. His underlying serious lung condition, however, made it conceivable that the narrowing was playing a role. Unfortunately, his lung condition also increased the risk of treatment.

I left the decision to him. He thought about it for a few days and decided to go for stenting his right coronary artery. I was a bit surprised, since over the years he had been so averse to any procedures and even medications. Remarkably, he felt energized right after the procedure was done. Because the stent was put in via the artery of his wrist, he went home just a few hours later. By that evening, he had walked several blocks and before the week's end he was swimming multiple laps. He told me he felt stronger and better than he had for several years. And, months later, the striking improvement in exercise capacity endured.

What's remarkable about this story is that a computer algorithm would have missed it. For all the hype about the use of AI to improve healthcare, had it been applied to this patient's data and the complete corpus of medical literature, it would have concluded not to do the procedure because there's no evidence that indicates the opening of a right coronary artery will alleviate symptoms of fatigue—and AI is capable of learning what to do only by examining existing evidence. And insurance companies using algorithms certainly would have denied reimbursement for the procedure.

But the patient manifested dramatic, sustained benefit. Was this a placebo response? That seems quite unlikely—I've known this man for many years, and he tends to minimize any change, positive or negative, in his health status. He seems a bit like a Larry David personality with curbed enthusiasm, something of a curmudgeon. Ostensibly, he would be the last person to exhibit a highly exaggerated placebo benefit.

In retrospect, the explanation likely does have something to do with his severe lung disease. Pulmonary fibrosis results in high pressures in the pulmonary arteries, which feed blood to the lungs, where the blood becomes oxygenated. The right ventricle is responsible for pumping that blood to the heart; the high blood pressure in the arteries meant that it would have taken a lot of work to force more blood in. That would have stressed the right ventricle; the stent in the right coronary artery, which supplies the right ventricle, would have alleviated the stress on this heart chamber. Such a complex interaction of one person's heart blood supply with a rare lung disease had no precedent in the medical literature.

This case reminds us that we're each a one-of-a-kind intricacy that will never be fully deconvoluted by machines. The case also highlights the human side of medicine: We physicians have long known that patients know their body and that we need to listen to them. Algorithms are cold, inhumane predictive tools that will never know a human being. Ultimately, this gentleman had a sense that his artery narrowing was the culprit for his symptoms, and he was right. I was skeptical and would certainly not have envisioned the magnitude of impact, but I was thrilled he improved.

<hr />

AI HAS BEEN sneaking into our lives. It is already pervasive in our daily experiences, ranging from autocomplete when we type, to unsolicited recommendations based on Google searches, to music suggestions based on our listening history, to Alexa answering questions or turning out the lights. Conceptually, its roots date back more than eighty years, and its name was coined in the 1950s, but only recently has its potential impact in healthcare garnered notice. The promise of artificial intelligence in medicine is to provide composite, panoramic views of individuals' medical data; to improve decision making; to avoid errors such as misdiagnosis and unnecessary procedures; to help in the ordering and interpretation of appropriate tests; and to recommend treatment. Underlying all of this is data. We're well into the era of Big Data now: the world produces zettabytes (sextillion bytes, or enough data to fill roughly a trillion smartphones) of data each year. For medicine, big datasets take the form of whole-genome sequences, high-resolution images, and continuous output from wearable sensors. While the data keeps pouring out, we've really processed only a tiny fraction of it. Most estimates are less than 5 percent, if that much. In a sense, it was all dressed up with nowhere to go—until now. Advances in artificial intelligence are taming the unbridled amalgamation of Big Data by putting it to work.

There are many subtypes of AI. Traditionally machine learning included logistic regression, Bayesian networks, Random Forests, support vector machines, expert systems, and many other tools for data analysis. For example, a Bayesian network is a model that provides probabilities. If I had a person's symptoms, for example, such a model could yield a list of possible diagnoses, with the probability of each one. Funny that in the 1990s, when we did classification and regression trees to let the data that we collected speak for itself, go into "auto-analyze" mode, without our bias of interpretation, we didn't use the term "machine learning." But now that form of statistics has undergone a major upgrade and achieved venerability. In recent years, AI tools have expanded to deep network models such as deep learning and reinforcement learning (we'll get into more depth in Chapter 4).

The AI subtype of deep learning has gained extraordinary momentum since 2012, when a now-classic paper was published on image recognition.[2]

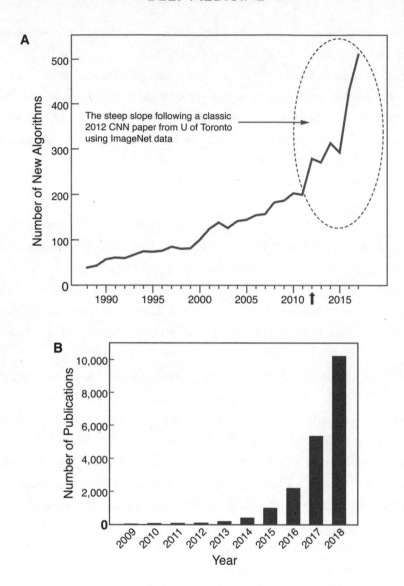

FIGURE 1.1: The increase in deep learning AI algorithms since the 2012 image recognition paper. Sources: Panel A adapted from A. Mislove, "To Understand Digital Advertising, Study Its Algorithms," *Economist* (2018): www.economist.com/science-and-technology /2018/03/22/to-understand-digital-advertising-study-its-algorithms. Panel B adapted from C. Mims, "Should Artificial Intelligence Copy the Human Brain?" *Wall Street Journal* (2018): www.wsj.com/articles/should-artificial-intelligence-copy-the-human-brain-1533355265 ?mod=searchresults&page=1&pos=1.

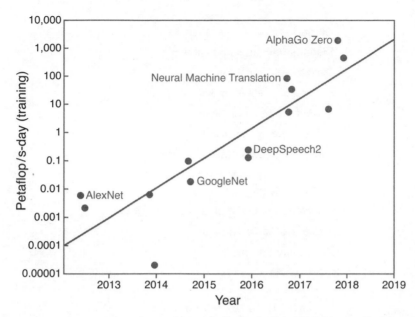

FIGURE 1.2: The exponential growth in computing—300,000-fold—in the largest AI training runs. Source: Adapted from D. Hernandez and D. Amodei, "AI and Compute," *OpenAI* (2018): https://blog.openai.com/ai-and-compute/.

The number of new deep learning AI algorithms and publications has exploded (Figure 1.1), with exponential growth of machine recognition of patterns from enormous datasets. The 300,000-fold increase in petaflops (computing speed equal to one thousand million million [10^{15}] floating-point operations per second) per day of computing used in AI training further reflects the change since 2012 (Figure 1.2).

In the past few years, several studies relying on deep learning have been published in leading peer-reviewed medical journals. Many in the medical community were frankly surprised by what deep learning could accomplish: studies that claim AI's ability to diagnose some types of skin cancer as well as or perhaps even better than board-certified dermatologists; to identify specific heart-rhythm abnormalities like cardiologists; to interpret medical scans or pathology slides as well as senior, highly qualified radiologists and pathologists, respectively; to diagnose various eye diseases as well as ophthalmologists;

Outperform doctors at all tasks

Diagnose the undiagnosable

Treat the untreatable

See the unseeable on scans, slides

Predict the unpredictable

Classify the unclassifiable

Eliminate workflow inefficiencies

Eliminate hospital admissions and
    readmissions

Eliminate the surfeit of unnecessary jobs

100% medication adherence

Zero patient harm

Cure cancer

TABLE 1.1: The outlandish expectations for AI in healthcare, a partial list.

and to predict suicide better than mental health professionals. These skills predominantly involve pattern recognition, with machines learning those patterns after training on hundreds of thousands, and soon enough millions, of examples. Such systems have just gotten better and better, with the error rates for learning from text-, speech-, and image-based data dropping well below 5 percent, whizzing past the human threshold (Figure 1.3). Although there must be some limit at which the learning stops, we haven't reached it yet. And, unlike humans who get tired, have bad days, may get emotional, sleep deprived, or distracted, machines are steady, can work 24/7 without vacations, and don't complain (although both can get sick). Understandably, this has raised questions about the future role of doctors and what unforeseen impact AI will have on the practice of medicine.

I don't believe that deep learning AI is going to fix all the ailments of modern healthcare, but the list in Table 1.1 gives a sense of how widely the tool can be applied and has been hyped. Over time,

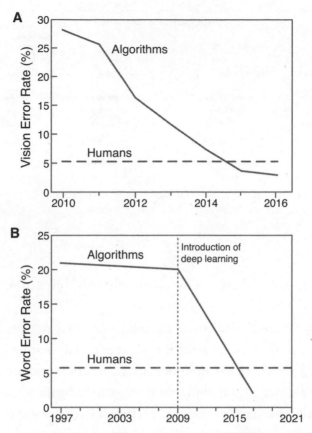

FIGURE 1.3: The increase in machine AI accuracy for image (A) and speech (B) interpretation, both now exceeding human performance for narrow tasks in labeled datasets. Sources: Panel A adapted from V. Sze et al., "Efficient Processing of Deep Neural Networks: A Tutorial and Survey," *Proceedings of the IEEE* (2017): 105(12), 2295–2329. Panel B adapted from "Performance Trends in AI," *Word Press Blog* (2018): https://srconstantin.wordpress.com/2017/01/28/performance-trends-in-ai/.

AI will help propel us toward each of these objectives, but it's going to be a marathon without a finish line.

The deep learning examples are narrow: the depression predictor can't do dermatology. These neural network algorithms depend on recognizing patterns, which is well-suited for certain types of doctors who heavily depend on images, like radiologists looking at scans or pathologists reviewing slides, which I call "doctors with patterns." To a lesser but still significant extent, all clinicians have some patterned

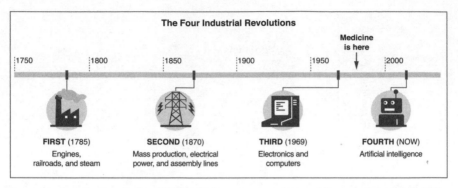

FIGURE 1.4: The four Industrial Revolutions. Source: Adapted from A. Murray, "CEOs: The Revolution Is Coming," *Fortune* (2016): http://fortune.com/2016/03/08/davos-new-industrial-revolution.

tasks in their daily mix that will potentially be subject to AI algorithmic support.

Most of the published deep learning examples represent only *in silico,* or computer-based, validation (as compared to *prospective* clinical trials in people). This is an important distinction because analyzing an existing dataset is quite different from collecting data in a real clinical environment. The in silico, retrospective results often represent the rosy best-case scenario, not fully replicated via a forward-looking assessment. The data from retrospective studies are well suited for generating a hypothesis, then the hypothesis can be tested prospectively and supported, especially when independently replicated.

We're early in the AI medicine era; it's not routine medical practice, and some call it "Silicon Valley–dation." Such dismissive attitudes are common in medicine, making change in the field glacial. The result here is that although most sectors of the world are well into the Fourth Industrial Revolution, which is centered on the use of AI, medicine is still stuck in the early phase of the third, which saw the first widespread use of computers and electronics (Figure 1.4). That MP3 files are compatible with every brand of music player, for example, while medicine has yet to see widely compatible and user-friendly electronic medical records exemplifies the field's struggle to change.

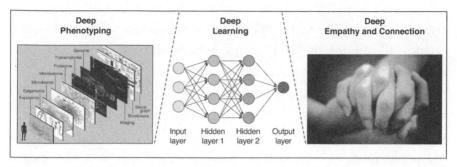

FIGURE 1.5: The three principal components of the deep medicine model. Source (left panel): Adapted from E. Topol, "Individualized Medicine from Prewomb to Tomb," *Cell* (2014): 157(1), 241–253.

This isn't the first time I've noted medicine's reluctance to adopt new technologies. This is the third book that I've written on the future of medicine. In the *Creative Destruction of Medicine,* I mapped out how sensors, sequencing, imaging, telemedicine, and many other technological opportunities enabled us to digitize human beings and achieve the digital transformation of medicine. In *The Patient Will See You Now,* I made the case for how medicine could be democratized—that medical paternalism would fade as consumers didn't simply generate their information but owned it, had far greater access to their medical data, and ultimately could (if they chose to) take considerably more charge of their care.

This book represents the next phase, the third *D* after digitizing and democratizing, and it's the most far-reaching one. Despite whatever impression you might get from my interest in new technology, it has always been a dream of mine to galvanize the essential human element of medical practice. With this third *D* of deep learning, we will have a framework to nurture medicine's roots: the human-human bond. Although we haven't even yet achieved digitization or democratization in medicine, they are slowly progressing, and I believe that we will not just complete them but bring AI into the heart of medicine as well. The culmination of this process is something I call "deep medicine."

Deep medicine requires three deep components (Figure 1.5).

First is the ability to deeply define each individual (digitizing the medical essence of a human being), using all relevant data. This might include all of one's medical, social, behavioral, and family histories, as well as one's biology: anatomy, physiology, and environment. Our biology has multiple layers—our DNA genome, our RNA, proteins, metabolites, immunome, microbiome, epigenome, and more. In the biomedical research community, the term that is frequently used is "deep phenotyping"; we saw an example of the approach in the case of the newborn boy with status epilepticus. Deep phenotyping is both thick, spanning as many types of data as you can imagine, and long, covering as much of our lives as we can, because many metrics of interest are dynamic, constantly changing over time. A few years ago, I wrote a review in which I said we needed medical data that spanned "from prewomb to tomb."[3] A former mentor told me that I should have called the span "from lust to dust." But you get the idea of deep and long data.

Second is deep learning, which will play a big part of medicine's future. It will not only involve pattern recognition and machine learning that doctors will use for diagnosis but a wide range of applications, such as virtual medical coaches to guide consumers to better manage their health or medical condition. It will also take on efficiency in the hospital setting, using machine vision to improve patient safety and quality, ultimately reducing the need for hospital rooms by facilitating remote, at-home monitoring. Although deep learning's outputs in medicine have considerable potential and have been accelerating in the past few years, we're still in the nascent phase. Nearly fifty years ago, William Schwartz published an article in the *New England Journal of Medicine* titled "Medicine and the Computer."[4] He speculated that, in the future, computers and physicians would engage "in frequent dialogue, the computer continuously taking note of history, physical findings, laboratory data, and the like, alerting the physician to the most probable diagnoses and suggesting the appropriate, safest course of action." What do we have to show for this projection from fifty years ago? Surprisingly, not too much. While there are cer-

METRIC	1975	NOW
Number of healthcare jobs	4 million	> 16 million (#1 US economy)
Healthcare spending per person	$550/yr.	> $11,000/yr.
Time allotted for office visits	60 min. new, 30 min. return	12 min. new, 7 min. return
% of GDP healthcare	< 8	18
Hospital daily room charge (avg.)	~ $100	$4,600
Miscellaneous	None of these	Relative value units, EHRs, PBMs, "health systems"

TABLE 1.2: Selected metrics of healthcare in the United States that have changed in the past forty-plus years.

tainly anecdotes about a Google search helping make difficult diagnoses, simple symptom lookup certainly has not been validated as an accurate means of diagnosis—instead, all too often, it serves as the groundwork for inducing anxiety and cyberchondria.

One can imagine that AI will rescue medicine from all that ails it, including diagnostic inaccuracy and workflow inefficiencies (such as mundane tasks like billing or coding charts), but none of these have been actualized yet. It's an extraordinary opportunity for entrepreneurs working with clinicians, computer scientists, and researchers in other disciplines (such as behavioral science and bioethics) to help fashion the right integration of AI and healthcare.

The third, and most important, component is deep empathy and connection between patients and clinicians. In the more than four decades since I started medical school, I've watched the steady degradation of the human side of medicine, outlined in Table 1.2. Over that span of time, healthcare became not just a big business but, by the end of 2017, the biggest. It is now the largest employer in the United States, towering over retail. By every metric, the amount of money spent on healthcare has exploded. Yet, even with all the employment in the sector and all the money expended per person, the time spent between doctors and patients has steadily dwindled, whether for office visits or in the hospital. Doctors are much too busy. The exorbitant charge of almost $5,000 for a day in the hospital might only include a few minutes of your doctor coming by to visit (for which there's another charge). Consumed by patient care, physicians were passive while major new changes took hold in the business of healthcare, including electronic health records, managed care, health maintenance organizations, and relative value units. Now, the highest-ever proportion of doctors and nurses are experiencing burnout and depression owing to their inability to provide real care to patients, which was their basis for pursuing a medical career.

What's wrong in healthcare today is that it's missing care. That is, we generally, as doctors, don't get to really care for patients enough. And patients don't feel they are cared for. As Francis Peabody wrote in 1927, "The secret of the care of the patient is caring for the patient."[5] The greatest opportunity offered by AI is not reducing errors or workloads, or even curing cancer: it is the opportunity to restore the precious and time-honored connection and trust—the human touch—between patients and doctors. Not only would we have more time to come together, enabling far deeper communication and compassion, but also we would be able to revamp how we select and train doctors. We have prized "brilliant" doctors for decades, but the rise of machines will heighten the diagnostic skills and the fund of medical knowledge available to all clinicians. Eventually, doctors will adopt AI and algorithms as their work partners. This leveling of the medical knowledge landscape will ultimately lead to a new premium: to find and train doctors who have the highest level of emotional intelligence. My friend and colleague, Abraham Verghese, whom I regard as one of the great humanists of medicine, has emphasized these critical points in the foreword of the book, which I hope you have taken the time to read carefully. This is what deep medicine offers.

To DEVELOP THE conceptual framework of deep medicine, I'll start with how medicine is practiced now and why we desperately need new solutions to such problems as misdiagnosis, errors, poor outcomes, and runaway costs. That, in part, hinges on the basics of how a medical diagnosis is made today. To understand the reward and risk potential of AI, we will explore the AI precedents, the accomplishments ranging from games to self-driving cars. Of equal, and perhaps even greater, importance will be an exploration of AI's liabilities, such as human bias, the potential for worsening inequities, its black-box nature, and concerns for breaches of privacy and security. The transfer of tens of millions of people's personal data from Facebook to Cambridge Analytica, who then used AI to target individuals, illustrates one critical aspect of what could go wrong in the healthcare context.

Then we're ready to move on to the new medicine that will integrate the tools of AI. We'll assess how machine pattern recognition will affect the practice of radiologists, pathologists, and dermatologists—the doctors with patterns. But AI will cut across all disciplines of medicine, even "clinicians without patterns" and surgeons. One field that is especially in urgent need of new approaches is mental health, with a profound mismatch of the enormous burden of conditions like depression and the limited number of trained professionals to help manage or prevent it. AI will likely prove to have a critical role in mental health going forward.

But AI, and specifically deep learning, won't just affect the practice of medicine. In a complementary way, it will also transform biomedical science. For example, it will facilitate the discovery of new drugs. It will also extract insights from complex datasets, such as millions of whole genome sequences, the intricacies of the human brain, or the integrated streaming of real-time analytics from multiple biosensor outputs. These endeavors are upstream from the care of patients, but catalyzing advances in basic science and drug development will ultimately have a major effect in medicine.

AI can also revolutionize other aspects of our lives that are, in one sense or another, upstream from the clinic. A huge one is how we eat. One of the unexpected and practical accomplishments of machine learning to date has been to provide a potential scientific basis for individualized diets. That's conceivably an exciting advance—the idea of knowing what specific foods are best for any given person. We can now predict in healthy people, without diabetes, what particular foods will spike their blood sugar. Such advances far outstrip whatever benefits might accrue from following a diet for all people, such as the classic food pyramids, or fad diets like Atkins or South Beach, none of which ever had a solid evidence basis. We'll review that fascinating body of data and forecast where smart nutrition may go in the future. Many of these at-home advances will come together in the virtual medical coach. It most likely will be voice mediated, like Siri, Alexa, and Google Home, but unlikely to remain a cylinder or a squiggle on a screen. I suspect they're more apt to come in the form of a virtual human avatar or hologram (but simply text or e-mail if one prefers). The virtual medical coach is the deep learning of all of one's data, seamlessly collected, continuously updated, integrated with all biomedical knowledge, and providing feedback and coaching. Such systems will initially be condition specific, say for diabetes or high blood pressure, but eventually they'll offer a broad consumer health platform to help prevent or better manage diseases.

All this potential, however, could be spoiled by misuse of your data. This encompasses not just the crimes we've seen all too much of so far, such as cybertheft, extortion (hospitals having their data held for ransom), and hacking, but the nefarious and large-scale sale and use of your data. The new, worrisome, unacceptable wrinkle could be that insurance companies or employers get hold of all your data— and what has been deep learned about you—to make vital decisions regarding your health coverage, your premiums, or your job. Avoiding such dreadful scenarios will take deliberate and intense effort.

This book is all about finding the right balance of the patients, doctors, and machines. If we can do that—if we can exploit machines' unique strengths to foster an improved bond between humans—we'll have found a vital remedy for what profoundly ails our medicine of today.

I hope to convince you that deep medicine is both possible and highly desirable. Combining the power of humans and machines—intelligence both human and artificial—would take medicine to an unprecedented level. There are plenty of obstacles, as we'll see. The path won't be easy, and the end is a long way off. But with the right guard rails, medicine can get there. The increased efficiency and workflow could either be used to squeeze clinicians more, or the gift of time could be turned back to patients—to use the future to bring back the past. The latter objective will require human activism, especially among clinicians, to stand up for the best interest of patients. Like the teenage students of Parkland rallying against gun violence, medical professionals need to be prepared to fight against some powerful vested interests, to not blow this opportunity to stand up for the primacy of patient care, as has been the case all too often in the past. The rise of machines has to be accompanied by heightened humaneness—with more time together, compassion, and tenderness—to make the "care" in healthcare real. To restore and promote care. Period.

Let's get started.

chapter two

SHALLOW MEDICINE

Imagine if a doctor can get all the information she needs about patient in 2 minutes and then spend the next 13 minutes of a 15-minute office visit talking with the patient, instead of spending 13 minutes looking for information and 2 minutes talking with the patient.

—Lynda Chin

"He told me I need a procedure to plug the hole in my heart," my patient, whom I'll call Robert, said at the beginning of our first encounter. Robert is a fifty-six-year-old store manager who had been healthy until a few years ago, when he had a heart attack. Fortunately, he received timely treatment with a stent, and there was very little damage to his heart. Since that time, he had markedly improved his lifestyle, losing and keeping off more than twenty-five pounds while exercising regularly and rigorously.

So, it was devastating for him when, out of the blue one afternoon, he began to have trouble seeing and developed numbness in his face. During an evaluation in the emergency room of a nearby hospital, the symptoms continued while he had an urgent head CT scan, some blood tests, a chest X-ray, and an electrocardiogram. Over the course of the day, without any treatment, his vision gradually returned to normal, and the numbness went away. The doctors told him that he had suffered "just" a ministroke, or transient ischemic attack, and that he should continue taking an aspirin each day as he had been since the heart attack. The lack of any change in strategy or new medication left him feeling vulnerable to another event. He set up an appointment with a neurologist for a couple of weeks later. Robert thought maybe that way he would get to the bottom of the problem.

The neurologist did some additional tests, including an MRI of his brain and an ultrasound evaluation of the carotid arteries in his neck, but he did not find anything to explain the transient stroke. He referred Robert to a cardiologist. The heart doctor did an echocardiogram that showed a patent foramen ovale (PFO). That's a tiny hole in the wall that separates the heart's two atria, the collecting chambers. Present in all fetuses (because it keeps blood from flowing to the lungs before we need to breathe), it typically closes when we take our first breaths; nevertheless, it remains open in about 15 to 20 percent of adults. "A-ha!" the cardiologist exclaimed to Robert. "This echo cinched the diagnosis." The cardiologist thought that a blood clot must have moved across the heart chambers and trekked up to his brain, where it caused the ministroke. To avoid any future strokes, he said, Robert needed to have a procedure to plug up that hole. It was scheduled for ten days later.

Well, Robert wasn't so sure about this explanation or the need for the procedure. He spoke to a mutual friend and soon came to see me for a second opinion. I was alarmed. Robert's PFO anatomy was far too common to be the definitive cause of the stroke based on such a minimal evaluation. Before invoking the hole as the cause of the stroke, a physician needs to exclude every other diagnosis. Plenty of people have such holes in their hearts and strokes, and they're not at all connected. If they were, a lot more of the one in five of us with PFOs would be suffering strokes. Furthermore, multiple randomized trials have tested the effectiveness of the treatment for cryptogenic strokes so-called because they have no known cause. Although these trials showed a consistent reduction in the number of subsequent stroke events, the implant and procedure lead to enough complications that the net benefit was marginal. And for Robert that was even more questionable because he did not have a full stroke, and his evaluations were not extensive enough to force us to fall back on the cryptogenic, default diagnosis yet.

Together, he and I developed a plan to hunt for other possible causes of the ministroke. One very common cause is a heart-rhythm disorder known as atrial fibrillation. To investigate that possibility, I ordered an unobtrusive Band-Aid-like patch called a Zio (made by iRhythm) for Robert to wear on his chest for ten to fourteen days. A chip in the patch captures an electrocardiogram of every heartbeat during the period in which it is worn. Robert wore his for twelve days. A couple of weeks later I got the results. Sure enough, Robert had several, otherwise asymptomatic, bouts of atrial fibrillation during that time. He didn't have any other symptoms because the heart rate never got too fast, and a few of the episodes occurred while he was asleep. The atrial fibrillation was a much more likely cause of the ministroke than the hole in his heart. We could use a blood thinner to hopefully prevent future events, and there was no need to move ahead with plugging the hole. Yes, there was a small risk of bleeding complications from the new medicine, but the protection from a future stroke warranted that trade-off. Robert was relieved when we discussed the diagnosis, treatment, and prognosis.

I don't present Robert because we were able to nail down the likely diagnosis. Although his story has a happy ending, it also represents everything wrong with medicine today. His experience, from the emergency room through the first visit to a cardiologist, is what I call shallow medicine. Rather than an emotional connection between patients and doctors, we have an emotional breakdown, with disenchanted patients largely disconnected from burned-out, depressed doctors. At the same time, there is a systemic problem with mistaken and excessive diagnosis, both of which can result in significant economic waste and human harm. In fact, the deficiencies in the

patient-doctor relationship and errors in medical practice are inter-
dependent: the superficial contact with patients promotes incorrect
diagnoses and the reflexive ordering of tests or treatments that are
unnecessary or unsound.

Misdiagnosis in the United States is disconcertingly common.
A review of three very large studies concluded that there are about
12 million significant misdiagnoses a year.[1] These mistakes result
from numerous factors, including failing to order the right test,
misinterpreting a test that was performed, not establishing a proper
differential diagnosis, and missing an abnormal finding. In Robert's
case, there was clearly a combination of errors: an incomplete differential diagnosis (possible atrial fibrillation), a failure to order the
right test (the heart-rhythm monitoring), and a misinterpretation of
the echocardiogram (attributing causal status to the PFO). A triple
whammy.

But the situation in the United States is worse than that because
misdiagnosis leads to mistreatment; Robert, for example, was set to
have a permanent implant to plug up the hole in his heart. In the
past few years much has been written about such unnecessary medical procedures. Shockingly, up to one-third of medical operations
performed are unnecessary.

Two big initiatives have been attempted to tackle this issue. The
first, called Choosing Wisely, began in 2012. The American Board
of Internal Medicine Foundation (ABIMF) worked with nine professional medical organizations to publish a list, titled "Five Things
Physicians and Patients Should Question," of the five most overused
or unnecessary tests and procedures.[2] Although the various medical
organizations had initially been reluctant to participate, the campaign
gained momentum over the next couple of years. Eventually more
than fifty medical societies joined, together identifying hundreds of
procedures and tests that provided low value for patients, given the
attendant costs or risks. By far, the most overused tests were medical
imaging studies for fairly innocuous conditions, such as lower back
pain or headache. To help bring this point home, for every one hun-

dred Medicare recipients age sixty-five or older, each year there are more than fifty CT scans, fifty ultrasounds, fifteen MRIs, and ten PET scans. It's estimated that 30 to 50 percent of the 80 million CT scans in the United States are unnecessary.[3]

While it was a coup to get the medical societies to fess up about their top five (and often top ten) misused procedures, there was ultimately little to show for the effort. Subsequent evaluation from a national sample showed that the top seven low-value procedures were still being used regularly and unnecessarily. Two primary factors seem to account for this failure. The first reason, called the therapeutic illusion by Dr. David Casarett of the University of Pennsylvania, was the established fact that, overall, individual physicians overestimate the benefits of what they themselves do.[4] Physicians typically succumb to confirmation bias—because they already believe that the procedures and tests they order will have the desired benefit, they continue to believe it after the procedures are done, even when there is no objective evidence to be found. The second reason was the lack of any mechanism to affect change in physicians' behavior. Although Choosing Wisely partnered with *Consumer Reports* to disseminate the lists in print and online, there was little public awareness of the long list of recommendations, so there was no grassroots, patient-driven demand for better, smarter testing. Furthermore, the ABIMF had no ability to track which doctors order what procedures and why, so there was no means to reward physicians for ordering fewer unnecessary procedures, nor one to penalize physicians for performing more.

In 2017 the RightCare Alliance, an international project organized by the Lown Institute in Boston, made a second attempt at reform. It published a series of major papers in the *Lancet* that quantified unnecessary procedures in a number of countries.[5] The United States was the worst offender, at as much as 60 percent. Again, medical imaging for conditions like back pain was at the top of the list. RightCare also examined some procedures that were appropriate but underused, although that problem pales in comparison. Much like Choosing Wisely was intended to shape physician behavior, the RightCare

Alliance hoped its extensive data would be incorporated into future medical practice. There are no data to suggest that has happened.

So that leaves us stuck where we have been—physicians regularly fail to choose wisely or provide the right care for patients. David Epstein of ProPublica wrote a masterful 2017 essay, "When Evidence Says No, But Doctors Say Yes," on the subject.[6] One example used in the article was stenting arteries for certain patients with heart disease: "Stents for stable patients prevent zero heart attacks and extend the lives of patients a grand total of none at all." As Epstein concluded about stenting and many other surgeries: "The results of these studies do not prove that the surgery is useless, but rather that it is performed on a huge number of people who are unlikely to get any benefit." Part of the problem is treatment that flies in the face of evidence, but another part of the problem is the evidence used to make the decisions to treat. In medicine, we often rely on changes in the frequency of so-called surrogate endpoints instead of the frequency of endpoints that really matter. So with heart disease we might treat based on changes in blood pressure because we have no evidence about whether the treatment actually changes the frequency of heart attacks, strokes, or death. Or we might measure the treatment of diabetes by monitoring changes in glycosylated hemoglobin (A1c) instead of life expectancy and widely accepted quality-of-life measures. Although the surrogate symptoms might seem like reasonable stand-ins for the overarching goals, very few of them have held up to scrutiny. Nevertheless, the flimsy evidence that caused physicians to monitor the surrogates in the first place has bred overuse of tests, procedures, and medications.

Shallow evidence, either obtained from inadequate examination of an individual patient like Robert, or from the body of medical literature, leads to shallow medical practice, with plenty of misdiagnoses and unnecessary procedures. This is not a minor problem. In 2017 the American Heart Association and American College of Cardiology changed the definition of high blood pressure, for example, leading to the diagnosis of more than 30 million more Americans

Figure 2.1: Word cloud describing doctors. Source: Adapted from B. Singletary et al., "Patient Perceptions About Their Physician in 2 Words: The Good, the Bad, and the Ugly," *JAMA Surg* (2017): 152(12), 1169–1170.

with hypertension despite the lack of any solid evidence to back up this guideline.[7] This was misdiagnosis at an epidemic scale.

Yet even without such central dictates, the way medical practice works at a one-to-one level is a setup for misdiagnosis. The average length of a clinic visit in the United States for an established patient is seven minutes; for a new patient, twelve minutes. This preposterous lack of time is not confined to America. When I visited the Samsung Medical Center in South Korea a couple of years ago, my hosts told me that the doctor visits averaged only two minutes. Is it any wonder that there are so many mistaken diagnoses? Both patients and doctors believe that doctors are rushed. Recently, for example, the medical center at the University of Alabama at Birmingham asked patients what two words best describe its doctors.[8] The response, plotted as a word cloud in Figure 2.1, is telling.

It's not just the length of a visit. Because of electronic health records, eye contact between the patient and doctor is limited. Russell Phillips, a Harvard physician said, "The electronic medical record has turned physicians into data entry technicians."[9] Attending to the keyboard, instead of the patient, is ascribed as a principal reason for the medical profession's high rates of depression and burnout. Nearly half of doctors practicing in the United States today have symptoms of burnout, and there are hundreds of suicides per year.[10] In a recent analysis of forty-seven studies involving 42,000 physicians, burnout was associated with a doubling of risk of patient safety incidents, which sets up a vicious cycle of more burnout and depression.[11] Abraham Verghese nailed this in the book's foreword, the role of the "intruder" and its impact on doctors' mental health, which beyond clinicians has potential impact on patient care.

The use of electronic healthcare records leads to other problems. The information that they contain is often remarkably incomplete and inaccurate. Electronic records are very clunky to use, and most—an average of 80 percent—of each note is simply copied and pasted from a previous note.[12] Any mistakes made on one visit are very likely to be propagated to the next. And getting records from other doctors and health systems is exceptionally difficult, in part because of proprietary issues: software companies use file formats that do not work on competitors' software, and health systems take advantage of proprietary file formats to help lock patients in. As my radiologist friend, Saurabh Jha, aptly put it on Twitter: "Your ATM card works in Outer Mongolia, but your electronic health record can't be used in a different hospital across the street."[13]

The incompleteness of the records is accentuated by one-off medicine. By "one-off," I'm not just referring to the brevity or rarity of the interaction. We haven't had access to patients in their real world, on the go, at work, while asleep. The data doctors access are from the contrived setting of a medical office, constrained by the temporal limits of the visit itself. A patient wearing a patch like I had Robert do is exceedingly rare. For the most part, we have no idea of what any given individual's real-life medical metrics—such as blood pressure, heart rate and rhythm, or level of anxiety and mood—actually are. In fact, even if we did know this for someone, we wouldn't have a means to make useful comparisons, as we don't even yet know what is normal for the population as a whole in a real-world context.

This is worsened by the outmoded means of communication doctors do—or don't—use to communicate with patients outside the clinic. Outside of medicine, people have learned how to maintain a close relationship with their family and friends through e-mail, texting, and video chats, even when they are in remote parts of the world. But more than two-thirds of doctors still do not leverage digital communication to augment their relationship with patients. The unwillingness to e-mail or text has been attributed to lack of time, medicolegal concerns, and lack of reimbursement, but I see them as another example of physicians' thin connection with their patients.

This is where we are today: patients exist in a world of insufficient data, insufficient time, insufficient context, and insufficient presence. Or, as I say, a world of shallow medicine.

THE OUTGROWTHS OF shallow medicine are waste and harm. Let's take the example of medical screening today. In the United States, mammography is recommended annually for women in their fifties. The total cost of the screening alone is more than $10 billion per year. Worse, if we consider 10,000 women in their fifties who have mammography each year for ten years, only five (0.05 percent) avoid a breast cancer death, while more than 6,000 (60 percent) will have at least one false positive result.[14] The latter might result in the harm and expense of a number of unnecessary procedures, including biopsies, surgery, radiation, or chemotherapy; at the very least, it results in considerable fear and anxiety.

Remarkably parallel to mammography is the use of prostate-specific antigen (PSA) screening for prostate cancer in men. Despite the American Urological Association's 2013 recommendation against the routine use of PSA screening, it is still widely practiced. Each year about 30 million American men are screened, 6 million have an elevated PSA, and 1 million have prostate biopsies. Approximately 180,000, or 18 percent, do get a diagnosis of prostate cancer, but an equal number of men have prostate cancer that the biopsy missed.[15] Added to this issue is the well-established, but frequently ignored, fact that most prostate cancer is indolent and will never threaten a patient's life. Multiple studies have validated genomic markers of the tumor that indicate aggressiveness and increased propensity to spread, but this information is still not incorporated into clinical practice.[16] Overall, the result is that there will be one prostate cancer death averted per one thousand men screened.[17] If you're overly optimistic, you could conclude that this benefit is twice as much as mammography (0.5 per 1,000)! Another way to look at the data: a man is 120 to 240 times more likely to be misdiagnosed from an abnormal PSA and 40 to 80 times more likely to have unnecessary radiation therapy or surgery than to have his life saved.

Cancer screening exemplifies almost every problem with shallow medicine. Back in 1999, South Korea started a national screening program for many types of cancer. The program was free of charge or involved a nominal copay for people with above-average income, which meant that a great many people participated in it. One of the tests was an ultrasound of the thyroid. In just over a decade, the rate of thyroid cancer diagnosis increased fifteenfold, making it the most common form of cancer in South Korea, with more than 40,000 people carrying the diagnosis. This might sound like a victory, but it was a meaningless diagnosis; there was no change of outcomes, including no difference in mortality related to thyroid cancer in South Korea, despite the widespread detection.[18]

This thyroid cancer screening story was replicated in the United States. A decade ago there were ads to "check your neck" with text: "Thyroid cancer doesn't care how healthy you are. It can happen to anyone, including you. That's why it's the fastest growing cancer in the U.S."[19] That turned out to be a self-fulfilling prophecy, leading to a big spike in incidence, as seen in Figure 2.2. More than 80 percent of the people diagnosed underwent thyroid gland removal and had to take medication to replace the hormones that the thyroid normally produces; almost half had radiation therapy of their neck.

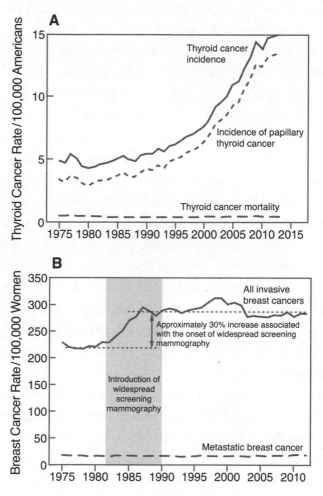

FIGURE 2.2: How mass screening leads to diagnoses without change in outcomes. Sources: Panel A adapted from H. Welch, "Cancer Screening, Overdiagnosis, and Regulatory Capture," *JAMA Intern Med* (2017): 177(7), 915–916. Panel B adapted from H. Welch et al., "Breast-Cancer Tumor Size, Overdiagnosis, and Mammography Screening Effectiveness," *N Engl J Med* (2016): 375(15), 1438–1447.

As was seen in South Korea, there was no sign that this aggressive diagnosis and treatment had any impact on outcomes. That's besides the danger of unnecessary radiation therapy itself.

Parallel to thyroid cancer, researchers at Dartmouth showed a very similar pattern with overdiagnosis of breast cancer (Figure 2.2).[20] Over the same period from 1975 to 2010, the new, routine use of mammography led to a 30 percent increase in the diagnosis of breast cancer, but over this same period there was no sign of reduced incidence of metastatic disease. For cancer, it's not the tumor per se but almost invariably the metastasis that kills—we now know that metastasis can occur quite early in the progression of a cancer. The mantra that early diagnosis of cancer would change the natural history and prevent bad outcomes was challenged.

For many decades, we have been taught in medical school that cancer takes many years to decades to form, through a slow process of doubling the tumor's cell populations that culminates in a mass, followed by another long phase before it becomes invasive and spreads to other parts of the body. That dogma has been seriously challenged by recent studies that show a tumor can spread in some patients at a very early phase of its development.[21] This inconvenient truth undermines a core tenet of screening, that early diagnosis of cancer will improve outcomes. It further exposes the problems of our predictive capacity in medicine for one of the leading causes of death and disability.

A great many of these problems could be avoided—and the tests and procedures could be done much more intelligently—if physicians actually took the time to identify whether a patient was at some risk of the disease they were trying to avoid. One important tool that has widespread awareness in medicine but is nevertheless regularly ignored is Bayes's theorem, which describes how knowledge about the conditions surrounding a possible event affect the probability that it happens. So, although we know about 12 percent of women will develop breast cancer during their lifetime, that does not mean that every woman has a 12 percent chance of developing breast cancer. For example, we know that people with certain BRCA gene mutations carry a very high risk, along with those who have a high genetic risk score. Screening all women without regard, for example, to a detailed family history (another outgrowth of lack of time with patients) or a screening for specific gene variants known to be related to breast cancer is a surefire scheme for generating false positives. For the same reason total body or MRI scans in healthy people yield an alarming number of incidental findings or, as Isaac Kohane nicknamed them, "incidentalomas."[22] Similarly, exercise stress tests in healthy people without symptoms lead to a high rate of abnormal results and prompt unnecessary angiograms. Many institutions across the United States cater to (and prey on) the fears of the healthy affluent with the mantra that early diagnosis of a condition could save their life. Many prestigious clinics do screenings for company executives, typically involving a large battery of unnecessary tests, with bills that range from $3,000 to $10,000. The chance of a false positive is multiplied by the number of unnecessary and groundless tests. Ironically, the subsequent workup that may be required for a false-positive, incidental finding may even put the patient's life at risk. Welch and colleagues, for example, documented that the unintended risk of an abdominal CT scan, which is done in 40 percent of Medicare beneficiaries within the first five years they are enrolled, is the increased chance they will be diagnosed with kidney cancer and have surgery to remove the kidney. That may sound absurd, but 4 percent of those patients die within ninety days from the surgery itself. What's more, there is no improvement in overall cancer survival in those who do survive the surgery.[23]

No test should be done on a willy-nilly, promiscuous basis, but rather its appropriateness should be gauged by the individual having some risk and suitability to be tested.

In the United States, we are now spending more than $3.5 trillion per year for healthcare. As seen in Table 2.1, for 2015, the number one line item is hospitals, accounting for almost a third of the costs.[24] The proportion attributable to doctors has remained relatively constant over many decades, at approximately one-fifth the costs. Prescription drugs are on a runaway course, accounting for well over $320 billion in 2015 and projected to reach over $600 billion by 2021.[25] New specialty drugs for cancer and rare diseases are routinely launched with price tags starting at $100,000 per treatment or year and ranging up to nearly $1 million per year.

CATEGORY	DOLLARS SPENT
Hospital care	1.0 trillion
Physician and clinical services	635 billion
Prescription drugs	325 billion
Next cost health insurance	210 billion
Nursing home and continuing care	157 billion
Dental services	118 billion
Structures and equipment	108 billion
Home healthcare	89 billion
Other professional services	88 billion
Government and public health activities	81 billion
Other durable medical products	59 billion
Research	47 billion
Government administration	43 billion

TABLE 2.1: Healthcare spending in the United States in 2015.

Part of this growth is fueled by a shared belief among both patients and physicians that medications, and in particular very expensive ones, will have remarkable efficacy. When doctors prescribe any medication, they have a cognitive bias that it will work. Patients, too, believe the medicine will work. From an enormous body of randomized clinical trials, patients assigned to the placebo arm consistently have more treatment effect than expected, given that they are taking an inert substance.

A few years ago, Nicholas Schork, a former faculty member at Scripps Research with me, put together the responsiveness—the intended clinical response—of the top ten drugs by gross sales.[26] As seen in Figure 2.3, the proportion of people who don't respond to these drugs is well beyond the common perception. Taking Abilify as an example, only one in five patients is actually deriving clinical benefit from the drug. Overall, 75 percent of patients receiving these leading medications do not have the desired or expected benefit. With several of these drugs with sales of more than $10 billion per year

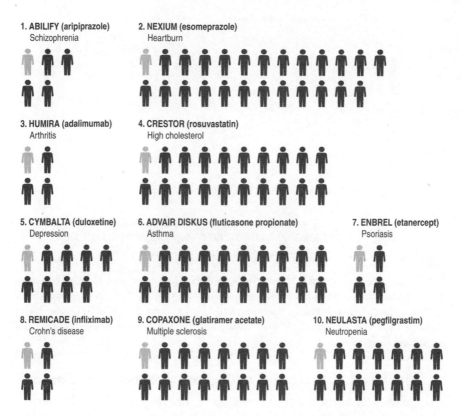

FIGURE 2.3: Schematic showing the number of people with clinical responsiveness to top ten drugs by gross sales in 2014. The gray schematic people represent clinical responders, the black nonresponders. Source: Adapted from N. Schork, "Personalized Medicine: Time for One-Person Trials," *Nature* (2015): 520(7549), 609–611.

(such as Humira, Enbrel, Remicade), you can quickly get a sense of the magnitude of waste incurred.

These data do not simply illustrate that medicines don't work or are some kind of profiteering racket. Rather, in most cases these drugs don't work because physicians have not honed an ability to predict what sort of person will respond to a treatment or acquired adequate knowledge about an individual to know whether the patient is among those people who will respond positively to a treatment. It adds to the continuum, from unintelligent diagnosis to treatment, of pervasive medical miscues, unnecessary interventions, and overuse problems that plague clinical practice today.

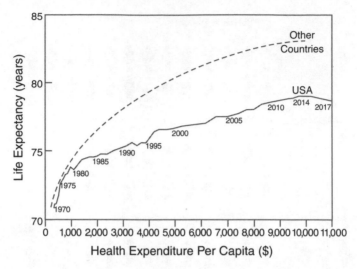

FIGURE 2.4: Life expectancy for twenty-four countries compared with the United States, plotted against health expenditures per person, from 1970 to 2017. Source: Adapted from M. Roser, "Link Between Health Spending and Life Expectancy: US Is an Outlier," *Our World in Data* (2017): https://ourworldindata.org/the-link-between-life-expectancy-and -health-spending-us-focus.

With all the unnecessary testing and treatment, misdiagnosis, and incidental findings that are chased down (and may do harm), we can look at perhaps the three most important measures of the efficacy of a healthcare system: longevity, infant/childhood mortality, and maternal mortality. They all look bad in the United States, and distinctly worse than the eighteen other member countries of the Organization for Economic Cooperation and Development (OECD) and beyond this group of countries (Figures 2.4 and 2.5). There are certainly other explanations that account for these outliers, such as the striking socioeconomic inequities in the United States, which continue to increase. For example, that appears to be a highly significant factor for the *alarming* and *disproportionate* maternal mortality rate among Black women.[27] I am not suggesting that the other countries are practicing deep medicine. In fact, my contention is that we, in the United States, are overindulged in shallow medicine. The evidence for overutilization, which is not the case for individuals of low socioeconomic status (who have problems with basic access), is

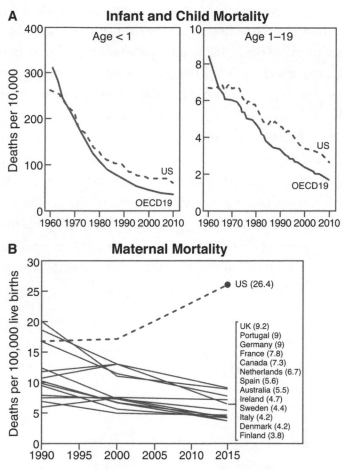

FIGURE 2.5: Reports showing the outlier status of the United States for key outcomes of (A) infant and child mortality and (B) maternal mortality. Sources: Panel A adapted from A. Thakrar et al., "Child Mortality in the US and 19 OECD Comparator Nations: A 50-Year Time-Trend Analysis," *Health Affairs* (2018): 37(1), 140–149. Panel B adapted from GBD Maternal Mortality Collaborators, "Global, Regional, and National Levels of Maternal Mortality, 1990–2015: A Systematic Analysis for the Global Burden of Disease Study 2015," *Lancet* (2016): 388(10053).

quite compelling and contributory. That our life expectancy is the singular one declining, while at the same time our healthcare spending is increasing, is deeply concerning.

For many years, healthcare economists have been talking about "bending the curve," meaning reducing costs to achieve the same or better outcomes. But with longevity declining in past years in the

United States, along with the continued steep increase in expenditures, we are indeed bending the curve, but in the wrong direction!

I hope that I've been able to convince you that the shallow medicine we practice today is resulting in extraordinary waste, suboptimal outcomes, and unnecessary harm. Shallow medicine is unintelligent medicine. This recognition is especially apropos in the information era, a time when we have the ability to generate and process seemingly unlimited data about and for any individual. To go deep. To go long and thick with our health data. That body of data—Big Data per individual—has the potential to promote the accuracy of diagnosis and treatment. We're not yet using it because it's far more than any human, any doctor can deal with. That's why we need to change the way we make medical diagnoses, the fundamental decision process of clinicians. Let's explore that issue next.

chapter three

MEDICAL DIAGNOSIS

> To be a good diagnostician, a physician needs to acquire a large set of labels for diseases, each of which binds an idea of the illness and its symptoms, possible antecedents and causes, possible developments and consequences, and possible interventions to cure or mitigate the illness.
>
> —DANIEL KAHNEMAN

> Computing science will probably exert its major effects by augmenting and, in some cases, largely replacing the intellectual functions of the physician.
>
> —WILLIAM B. SCHWARTZ, 1970

IT WAS THE BEGINNING OF MY THIRD YEAR OF MEDICAL SCHOOL. I was in the Introduction to Clinical Medicine clerkship at Strong Memorial Hospital in Rochester, New York. It was like the top of the first inning on opening day. The mentor for our group of ten students was Dr. Arthur Moss, a highly regarded cardiologist and teacher at the University of Rochester. (A mentor hero of mine, he passed away in 2018.) Before we went to a single patient's bedside, we were going to have a warm-up session in a conference room.

Moss's appearance was memorable: dark eyed with a bit of a squint when he looked at us, he had some silver strands in his jet-black hair. He wore a long white cloth-buttoned lab coat that seemed to extend too far below his knees, charcoal pants with cuffs, black socks, and black wing-tip shoes. His goal for this morning was to teach us the basics of how to make a diagnosis.

He went over to the chalkboard (there were no whiteboards in 1977) and started writing some features about a patient.

First he wrote, "Sixty-six-year-old male presents to the emergency room."

Then he asked, "What is in our differential diagnosis?"

Now this would seem peculiar since there was so little information to go on. But Dr. Moss's point was that every time a physician assesses a case, we need to process every nugget of information—be it a symptom, a sign, or a lab test result—and quickly come up with the most common causes that fit the picture.

The responses he got from our group of uninitiated doctor-wannabees were heart attack, cancer, stroke, and accident.

Then he added another feature: chest pain.

The group concluded that he must be having a heart attack.

Dr. Moss, looking at us wryly, told us that we were all wrong. We need to be thinking of other reasons for chest pain in such a patient. Then we came up with other possibilities like aortic dissection, esophageal spasm, pleurisy, pericarditis, and a contusion of the heart.

He now wrote on the board that the chest pain was radiating to the neck and the back. We zeroed in on heart attack and aortic dissection. Then he added that the patient had briefly passed out, which led to our final diagnosis of aortic dissection. Moss smiled and said, "Correct." He told us to never forget that possibility of aortic dissection when you see a patient with chest pain. That it was too often not considered and missing it could prove to be a fatal mistake.

Next it was time for a more difficult challenge. After erasing the board, he wrote, "Thirty-three-year-old woman is admitted to the hospital."

We responded with breast cancer, complication of pregnancy, accident. Moss was disappointed we didn't have more ideas to offer. The next feature he gave us was rash.

Our differential now extended to an infection, an adverse reaction to a drug, an insect or animal bite, a bad case of poison ivy. Our mentor, again looking a bit discouraged with us, had to add another feature to help us: facial rash. That didn't seem to get us on the right track. We were stuck with the same differential list. So, he added one more descriptor about his made-up patient: she was African American.

One of our group whispered, "Lupus?"

That was the right answer. She nailed it, knowing that lupus is much more common in young women of African ancestry and one of its hallmarks is a butterfly facial rash.

This is how we learned to make a medical diagnosis. It was top-down, immediately reacting to a few general descriptors and quickly coming up with a short list of hypotheses, conjectures, tentative conclusions. We were instilled with the mantra that common things occur commonly, an outgrowth of the same logic that underlies Bayes's theorem. We were getting programmed to use our intuitive sense of recognition rather than our analytical skills. But Bayes's theorem relies on priors, and, because we, as inexperienced medical students, had visited so many books but so few patients, we didn't have much to go on. The method would leave aged physicians, who had seen thousands of patients, in far better stead.

The diagnostic approach we were being taught is an example of what Danny Kahneman would one day classify as System 1 thinking—thinking that is automatic, quick, intuitive, effortless.[1] This system of thinking uses heuristics, or rules of thumb: the reflexive, mental shortcuts that bypass any analytic process, promoting rapid solutions to a problem. System 2 thinking, in contrast, is a slow, reflective process involving analytic effort. It occurs in a different area of the brain and even has distinct metabolic requirements. One might think that master diagnosticians would rely on System 2 thinking. But no, multiple studies have shown that their talent is tied to heuristics admixed with intuition, experience, and knowledge. Indeed, more than forty years ago, System 1 thinking, represented by the rapid, reflexive hypothesis generation method every physician is taught, was shown to be the prototype for getting the right diagnosis. If a doctor thought of the correct diagnosis within five minutes of seeing a patient, the accuracy was a stunning 98 percent. Without having the diagnosis in mind by five minutes, the final accuracy was only 25 percent.[2]

One medical environment stands out for the challenge—the emergency room, where physicians must assess each patient quickly and either admit them to the hospital or send them home. A wrong diagnosis can result in a person's death soon after being discharged, and, with almost 20 percent of the population of the United States visiting an emergency room each year, the population at risk is huge. A large study of ER evaluations of Medicare patients showed that each year more than 10,000 people died within a week of being sent home, despite not having a previously diagnosed illness or being diagnosed with a life-threatening one.[3] This isn't simply a problem in the emergency room. There are more than 12 million serious diagnostic errors each year in the United States alone,[4] and, according to a landmark report published in 2015 by the National Academy of Sciences, most people will experience at least one diagnostic error in their lifetime.[5]

These data point to the serious problems with how physicians diagnose. System 1—what I call fast medicine—is malfunctioning, and so many other of our habitual ways of making an accurate diagnosis can be improved. We could promote System 2 diagnostic reasoning. Kahneman has argued that "the way to block errors that originate in System 1 is simple in principle: recognize the signs that you are in a cognitive minefield, slow down and ask for reinforcement from System 2."[6] But to date, albeit with limited study, the idea that we can supplement System 1 with System 2 hasn't held up: when doctors have gone into analytic mode and consciously slowed down, diagnostic accuracy has not demonstrably improved.[7] A major factor is that the use of System 1 or System 2 thinking is not the only relevant variable; other issues come into play as well. One is a lack of emphasis on diagnostic skills in medical education. Of the twenty-two milestones of the American Board of Internal Medicine Accreditation Council for Graduate Medical Education, only two are related to diagnostic skills.[8] Once trained, doctors are pretty much wedged into their level of diagnostic performance throughout their career. Surprisingly, there is no system in place for doctors to get feedback on their diagnostic skills during their careers, either. In *Superforecasting,* Philip Tetlock observes, "If you don't get feedback, your confidence grows much faster than your accuracy."[9] The lack of emphasis on diagnostic skills during and after medical school, however, seems to be overshadowed by the lack of appreciation of deep cognitive biases and distortions that can lead to diagnostic failure. They're not even part of teaching diagnosis today in medical school.

In *The Undoing Project: A Friendship That Changed Our Minds,* Michael Lewis wrote about Donald Redelmeier, a Canadian physician who as a teenager was inspired by Amos Tversky and Danny Kahneman.[10] At Sunnybrook Hospital's trauma center, he asked his fellow physicians to slow down, tame System 1 thinking, and try to avoid mental errors in judgment. "You need to be so careful when there is one simple diagnosis that instantly pops into your mind that beautifully explains everything all at once. That's when you need to stop and check your thinking."[11] When a patient was misdiagnosed to be hyperthyroid for her irregular heartbeat but instead was found to have fractured ribs and a collapsed lung, Redelmeier called this error an example of the representativeness heuristic, which is a shortcut in decision making based on past experiences (first described by Tversky and Kahneman). Patterns of thinking such as the representativeness heuristic are an example of the widespread problem of cognitive bias among physicians. Humans in general are beset by many biases—*Wikipedia* lists 185, for example—but I want to highlight only a few of those that impair diagnostic accuracy.[12] It's important to emphasize that these embedded cognitive biases in medicine are simply human nature, not at all specific to making a diagnosis or being sure about recommending a treatment. But what is different here is that medical decision making can have profound, even life-and-death, consequences.

Some of the cognitive biases that lead to errors in diagnosis are quite predictable. There are about 10,000 human diseases, and there's not a doctor who could recall any significant fraction of them. If doctors can't remember a possible diagnosis when making up a differential, then they will diagnose according to the possibilities that are mentally "available" to them, and an error can result. This is called the availability bias.

A second bias results from the fact that doctors deal with patients one at a time. In 1990, Redelmeier and Tversky published a study in the *New England Journal of Medicine* that showed how individual patients, especially patients that a doctor has recently seen, can shape medical judgment, simply because each doctor only ever sees a relatively small number of patients.[13] Their personal experience as doctors can override hard data derived from much larger samples of people, say, about the likelihood that a patient has some rare disease, simply because an earlier patient with similar symptoms had that rare disease. Like when I saw a patient with a stroke who had a very rare tumor on a heart valve (called papillary fibroelastoma) and thought of it as a potential culprit in many subsequent patients. Compounding this is the fact that, as Redelmeier has found, 80 percent of doctors don't think probabilities apply to their patients.

One example of this bias from my experience comes to mind. Inserting a coronary stent has a small chance of inducing a heart attack in the patient. These heart attacks are rarely accompanied by any symptoms but can be diagnosed with blood test enzymes that prove there has been some damage to heart muscle cells. When my colleagues and I published a series of papers in the 1990s about this issue, known as periprocedural myocardial infarction, the reaction of most cardiologists was that we were wrong, that the problem was completely overblown. But each cardiologist was performing fewer than a hundred to a few hundred procedures per year, and they were not routinely checking blood tests to see whether there was any evidence of heart damage. And all the doctors were influenced by a bias to believe that they were highly skilled and so wouldn't be inducing heart attacks in their patients. Here the cognitive bias of doctors was influenced by their own relatively limited clinical experience and their failure to systematically look for evidence.

Rule-based thinking can also lead to bias. Cardiologists diagnosing heart disease in patients evaluated in the emergency department demonstrate such bias, as shown in Figure 3.1, when they assume that a patient must be over age forty before they really suspect heart attack. The evidence is clear, as shown by Stephen Coussens in his nicely titled paper, "Behaving Discretely: Heuristic Thinking in the Emergency Department": there is a discontinuity in the data (Figure 3.1A) that indicates doctors were classifying patients as too young to have heart disease, even though the actual risk of a forty-year-old having a fatal heart attack is not much greater than the risk of a thirty-nine-year-old patient having one (Figure 3.1B). This matters: having examined the ninety-day follow-up data for the patients in question, Coussens found that many individuals who were incorrectly deemed too young to have heart disease subsequently had a heart attack.[14]

One of the greatest biases prevalent among physicians is overconfidence, which Kahneman called "endemic in medicine."[15] To support his assertion, he recalls a study that determined physician confidence in their diagnoses and compared causes of death as ascertained by autopsy with the diagnoses the physicians had made before the patients died. "Clinicians who were 'completely certain' of the diagnosis antemortem were wrong 40 percent of the time." Lewis understood this bias, too: "The entire profession had arranged itself as if to confirm the wisdom of its decisions."[16] Tversky and Kahneman discussed a bias toward certainty in a classic 1974 paper in *Science* that enumerated the many types of heuristics that humans rely on when dealing with uncertainty.[17] Unfortunately, there has never been a lack of uncertainty in medicine, given the relative

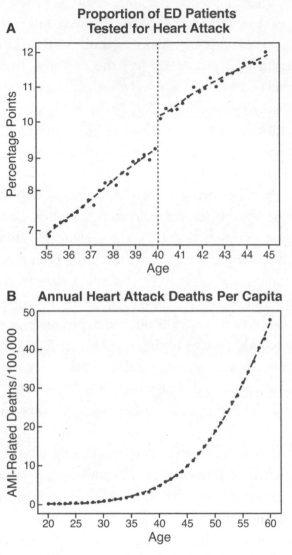

FIGURE 3.1: Heuristic thinking leads to misdiagnosis of heart attack in the emergency room. Source: Adapted from S. Coussens, "Behaving Discretely: Heuristic Thinking in the Emergency Department," *Harvard Scholar* (2017): http://scholar.harvard.edu/files/coussens /files/stephen_coussens_JMP.pdf.

dearth of evidence in almost every case. Unfortunately, dealing with that uncertainty often leads to a dependence on expert opinions, what I call eminence-based medicine (reviewed in depth in *The Creative Destruction of Medicine*).[18]

What Went Wrong

The leading causes of diagnostic errors in a sample of 583 physician-reported cases

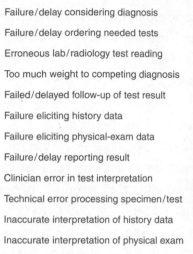
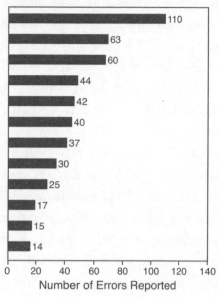

Failure/delay considering diagnosis	110
Failure/delay ordering needed tests	63
Erroneous lab/radiology test reading	60
Too much weight to competing diagnosis	44
Failed/delayed follow-up of test result	42
Failure eliciting history data	40
Failure eliciting physical-exam data	37
Failure/delay reporting result	30
Clinician error in test interpretation	25
Technical error processing specimen/test	17
Inaccurate interpretation of history data	15
Inaccurate interpretation of physical exam	14

Number of Errors Reported

FIGURE 3.2: The attributable causes of medical diagnostic errors in a sample of more than five hundred doctors. Source: Adapted from L. Landro, "The Key to Reducing Doctors' Misdiagnoses," *Wall Street Journal* (2017): www.wsj.com/articles/the-key-to-reducing-doctors-misdiagnoses-1505226691, with primary reference from G. Schiff et al., "Diagnostic Error in Medicine: Analysis of 583 Physician-Reported Errors," *Arch Intern Med* (2009): 169(20), 1881–1887.

Some of this overconfidence can be classified as confirmation bias, also known as myside bias—the tendency to embrace information that supports one's beliefs and to reject information that contradicts them.[19] Overconfidence is closely related to the illusion of explanatory depth, when humans believe they know more than they actually do. Whatever the type of bias, it's clear that humans (and that includes doctors) make key decisions that depart from rational behavior.

A classic experiment that Tversky performed further reinforces lack of simple reasoning. He did a survey of cancer doctors at Stanford and asked them to choose an operation for patients with terminal cancer. When given a choice described as having a 90 percent chance of survival, 82 percent chose it. But when it was described as a 10 percent chance of dying, only 54 percent selected the option. Just flipping the terms "survival" and "dying," and the corresponding percentages, led to a marked change in choice.

So, we know a lot about misdiagnosis, like approximately how many occur each year and that a significant proportion are due to cognitive biases. From a study of 583 doctor-reported cases of diagnostic errors, the biggest problem is not considering the diagnosis (Figure 3.2) in the first place, which is an outgrowth of System 1 thinking and availability bias.[20] Failure to diagnose or delay in diagnosis are the most important reasons for malpractice litigation in the United States, which in 2017 accounted for 31 percent of lawsuits.[21] When the affected physicians were asked what they would have done differently, the most common response was to have had better chart documentation, which again reflects the speed with which encounters and record keeping typically occur. Clearly, it is critical that the occurrence of misdiagnoses be markedly reduced, even as we must acknowledge that we will never get to zero.

Either shallow or fast medicine, by itself, is a significant problem. We've got to contend with them both. Even in the very uncommon situations when a physician's knowledge of a patient is fairly deep and a near-comprehensive dataset can be assembled, all the shortcomings of human thought and the limited scope of human experience come into play. Cumulatively, each physician may see thousands of patients during his or her career. This experience is the basis for each doctor's System 1 thinking, although, as I've mentioned, there is no mechanism in place for giving physicians regular feedback about whether they were correct. It takes many decades for each doctor to accumulate this experience, which, in reality, is quite limited; indeed, even for the outlier doctor who might see tens of thousands of patients over the course of a career, the number is small, compared with what could be amassed by aggregation of data tied with experience from large cohorts of doctors, like the more than 700,000 physicians currently practicing in the United States or several million worldwide. Enter computers.

A potential aid to physicians is online tools. Although there are certainly anecdotes about a Google search helping make difficult diagnoses, simple symptom lookup certainly has not been validated as an accurate means of diagnosis. One of the earliest symptom checkers used by doctors, and now patients, is the Isabel Symptom Checker, which covers more than 6,000 diseases. When I entered cough and fever for a fifty- to sixty-four-year-old male in North America, the "likely" diagnoses were influenza, lung cancer, acute appendicitis, lung abscess, relapsing fever, atypical pneumonia, and pulmonary embolism. It would be pretty easy to rule out almost all of these besides influenza and atypical pneumonia because of the likelihood of the symptoms being associated with these conditions. Back in 2015, a study published in the *British Medical Journal* assessed twenty-three symptom checkers. Only 34 percent had the correct diagnosis after the information was put into the system.[22] Despite that poor showing, in recent years mobile apps for checking symptoms, such as Ada, Your.MD, and Babylon, have proliferated. They incorporate components of artificial intelligence, but they haven't yet been shown to simulate the accuracy of diagnoses made by doctors (which we should not necessarily regard as the gold standard). These start-up companies are beginning to incorporate information beyond lists of symptoms, asking a series of questions like the patient's health history. The back-and-forth queries are hoped to narrow the differential and promote accuracy. One such app, Buoy Health, draws upon more than 18,000 clinical publications, descriptions of 1,700 medical conditions, and data provided by more than 5 million patients.

Nevertheless, the idea that a cluster of symptoms can lead to a correct diagnosis seems oversimplified. When listening to patients, it's abundantly clear that presence of a symptom is not 0 or 1, binary; rather, symptoms are nuanced and colored. For example, a patient suffering aortic dissection might not describe the sensation as "chest pain." For a heart attack, the patient could show his or her clenched fist (known as Levine's sign), denoting a sense of pressure not perceived as pain.

Or it could be a burning that isn't felt as either pressure or pain. Further complicating the matter for such diagnostic applications is that not only are symptoms subjective, how they are conveyed by the patient via descriptors, facial expression, and body language is critical and often cannot be readily captured by a few words.

Computers can also aid with getting a second opinion, which can improve the likelihood of arriving at the correct diagnosis. In a Mayo Clinic study that looked at nearly three hundred consecutive patients referred, the second opinion diagnosis agreed with the referral physician diagnosis in only 12 percent of patients.[23] Worse, second opinions often don't get done, owing, in part, to the cost, the difficulties in getting an appointment, or even finding an expert physician to turn to. Telemedicine makes getting input on an important diagnosis easier, although we are trading off the benefit of an in-person consultation against the benefit of having an additional doctor weigh in. When I was at the Cleveland Clinic at the turn of the millennium, we started an online service called MyConsult, which has now provided tens of thousands of second opinions, often leading to disagreement with the original diagnosis.

Doctors hoping to facilitate diagnostic accuracy can crowdsource data with their peers and find help with diagnostic work. This isn't exactly System 2 thinking, but taking advantage of the reflexive input and experience of multiple specialists. In recent years several smartphone apps for doctors have cropped up, including Figure One, HealthTap, and DocCHIRP. Figure One, for example, is pretty popular for sharing medical images to get a quick diagnosis from peers. My team at Scripps recently published data from what is currently the most widely used doctor crowdsourcing app, Medscape Consult.[24] Within two years of launch, the app was used by a steadily growing population of 37,000 physicians, representing more than two hundred countries and many specialties, with rapid turnaround on requests for help; interestingly, the average age of users was more than sixty years. The Human Diagnosis Project, also known as Human Dx, is a web- and mobile-app-based platform that has been used by more than 6,000 doctors and trainees from forty countries.[25] In a study that compared more than 200 doctors with computer algorithms for reviewing a diagnostic vignette, the diagnostic accuracy for doctors was 84 percent but only 51 percent for algorithms. That's not too encouraging for either the doctors or AI, but the hope of its leaders, with backers from many organizations like the American Medical Association, American Board of Medical Specialties, and other top medical boards, is that the collective intelligence of doctors and machine learning will improve diagnostic accuracy. One anecdote from its leader, Dr. Shantanu Nundy, an internist, is optimistic.[26]

He was seeing a woman in her thirties with stiffness and joint pain in her hands. He was unsure of the diagnosis of rheumatoid arthritis, so he posted on the HumanDx app "35F with pain and joint stiffness in L/R hands X 6 months, suspected rheumatoid arthritis." He also uploaded a picture of her inflamed hands. Within hours, multiple rheumatologists confirmed the diagnosis. Human Dx intends to recruit at least 100,000 doctors by 2022 and increase the use of natural-language-processing algorithms to direct the key data to the appropriate specialists, combining AI tools with doctor crowdsourcing.

An alternative model for crowdsourcing to improve diagnosis incorporates citizen science. Developed by CrowdMed, the platform sets up a financially incentivized competition among doctors *and* lay people to crack difficult diagnostic cases. The use of non-clinicians for this purpose is quite novel and has already led to unexpected outcomes: as Jared Heyman, the company's founder and CEO told me, the lay participants have a higher rate of accurate diagnosis than the participating doctors do. Our team at Scripps Research hasn't had the opportunity to review their data or confirm the final diagnosis accuracy. But, if validated, an explanation for this advantage may be that lay people have more time to extensively research the cases, reinforcing the value of slowness and the depth of due diligence to get the right answer in difficult cases.

The one company that has boldly broadcast its ambitious plans to improve medical diagnoses and outcomes is IBM, through its Watson supercomputer and use of artificial intelligence (Figure 3.3). In 2013, IBM started working with leading medical centers; spending billions of dollars and buying companies; and feeding Watson patient data, medical images, patient histories, biomedical literature, and billing records.[27] By 2015 IBM claimed that Watson had ingested 15 million pages of medical content, more than two hundred medical textbooks, and three hundred medical journals. The vast number of publications that come out every day surely represents a knowledge base the medical profession is unable to keep up with but that is worth tapping. In 2016 IBM even acquired Truven Health Analytics for $2.6 billion so that it could bring in 100 million patient records that Truven had access to, exemplifying Watson's insatiable need for medical data.[28]

When the Watson team visited our group at Scripps Research a few years ago, it demonstrated how inputting symptoms would yield a differential diagnosis ranked by probabilities. But if we wanted to use it for our unknown disease genomic sequencing program, we would have to pony up more than $1 million. We couldn't afford to do that. But other centers were undeterred, and the reports have certainly been mixed.

The most glowing results, from the University of North Carolina's Lineberger Comprehensive Cancer Center, were featured on *60 Minutes* in 2016. The cancer center's director, Norman Sharpless (who is now the National Cancer Institute director), described himself as a skeptic of the use of artificial intelligence to improve cancer outcomes. Of all the treatments considered for one thousand cancer patients at UNC, 30 percent were identified by Watson alone, on the basis of the system's analysis of the peer-reviewed literature for cancer.[29] Suggesting treatments is not the same thing as improving diagnoses, but Watson's ability to devour the 160,000-plus cancer research papers published per year might be the ticket to helping some patients. The study's first peer-reviewed publication with more than

FIGURE 3.3: IBM Watson ad.

a thousand patients at the University of North Carolina identified suitable clinical trials that were initially missed by oncologists in more than three hundred people.[30]

IBM Watson's experience with MD Anderson, one of the country's leading cancer centers, was a debacle noteworthy for many missteps. One of the most fundamental was the claim that ingesting millions of pages of medical information was the same as being able to make sense, or use, of the information. As head of the MD Anderson project testing Watson, Dr. Lynda Chin put it, "Teaching a machine to read a medical record is harder than anyone thought."[31] It turns out that it's not so easy to get a machine to figure out unstructured data, acronyms, shorthand phrases, different writing styles, and human errors. That perspective was echoed by Dr. Mark Kris of Memorial Sloan Kettering, who was involved in early training of the system called Watson for Oncology: "Changing the system of cognitive computing doesn't turn around on a dime like that. You have to put in the literature, you have to put in cases."[32] Fragmentary clinical data and lack of evidence from the medical literature made this project of little value. Ultimately, the MD Anderson project with Watson cost $62 million, and it collapsed. It missed deadline after deadline, changing focus from one type of cancer to another, planning pilot projects that never got off the ground.[33] Perhaps it's not surprising that a former IBM manager, Peter Greulich, reflecting on the project, concluded,

"IBM ought to quit trying to cure cancer. They turned the marketing engine loose without controlling how to build and construct a product."[34] Noteworthy is the perspective of Isaac Kohane, who leads the Department of Biomedical Informatics at Harvard Medical School: "One of the top reported facts was that MD Anderson had created a platform for leukemia-based diagnoses, over 150 potential protocols, researchers use Watson to make platform, blah blah blah. It was never used, it didn't exist."[35]

The problems that IBM Watson encountered with cancer are representative of its efforts to improve diagnosis across medicine. A highly optimistic, very exuberant projection of Watson's future appears in *Homo Deus* by Yuval Noah Harari: "Alas, not even the most diligent doctor can remember all my previous ailments and check-ups. Similarly, no doctor can be familiar with every illness and drug, or read every new article published in every medical journal. To top it all, the doctor is sometimes tired or hungry or perhaps even sick, which affects her judgment. No wonder that doctors sometimes err in their diagnoses or recommend a less-than-optimal treatment."[36] There is certainly potential for computing to make a major difference, but so far there has been minimal delivery on the promise. The difficulty in assembly and aggregation of the data has been underestimated, not just by Watson but all tech companies getting involved with healthcare.

While we are nowhere near Harari's notion, we do need machine-assisted diagnosis. With the challenge of massive and ever-increasing data and information for each individual, no less the corpus of medical publications, it is essential that we upgrade diagnosis from an art to a digital data-driven science. Yet, so far, we have only limited prospective clinical trials to suggest this will ultimately be possible.

FAST AND NARROW

Until this point, we've focused on whole-patient diagnosis and not drilled down on more narrow aspects, like interpretation of medical scans, pathology slides, electrocardiograms, or voice and speech. It's in making sense of these patterns that machines are making substantial progress.

Let me provide a brief sampling of some of the progress of narrow AI diagnostics. For the brain, we've seen better accuracy of scan interpretations in patients who present with a stroke or the pickup of subtleties in brain images as a reliable tip-off for subsequent Alzheimer's disease. For heart studies, there has been accurate interpretation of electrocardiograms for heart-rhythm abnormalities and echocardiographic images. In cancer, machines have diagnosed skin lesions and pathology slides well. Much work has been done to advance diagnosis of many eye diseases quite accurately from retinal images. Processing of sounds—voice and speech—has helped in the diagnosis of post-traumatic stress disorder or traumatic brain injury. Even the audio waveforms of a cough have been used to assist the diagnosis of asthma, tuberculosis, pneumonia, and other lung conditions.

The Face2Gene app from FDNA is worth highlighting as it can help the diagnosis for more than 4,000 genetic conditions, many of which can be extremely difficult to pinpoint. An example is a child with the rare Coffin-Siris syndrome. The app makes the diagnosis by recognizing unique facial features in seconds, whereas in some families it has taken up to sixteen years of extensive and expensive evaluations for humans to make the same diagnosis. The app's creators accomplished this by applying deep learning to images of afflicted individuals, identifying the rare but distinctive constellation of facial features that are the hallmark of the syndrome. Already 60 percent of medical geneticists and genetic counselors have used the app. And that's especially good because its wide use continually expands the knowledge resource to more accurately diagnose an ever-increasing proportion of rare diseases. Here again you can see striking success of a narrow AI tool to improve medical diagnosis. But it's not just narrow. The machine processing can be remarkably fast and cheap. For medical imaging processing, it was estimated that more than 250 million scans could be read in twenty-four hours at the cost of about $1,000.[37]

This all sounds and looks promising, but it's remarkably super-ficial. To really understand the promise and expose the pitfalls, we need to go deep into AI technology. In this chapter, for example, I've talked a lot about human biases. But those same biases, as part of human culture, can become embedded into AI tools. Since progress in AI for medicine is way behind other fields, like self-driving cars, facial recognition, and games, we can learn from experience in those arenas to avoid similar mistakes. In the next two chapters, I'll build up and then take down the field. You'll be able to gain insights into how challenging it will be for AI to transform medicine, along with its eventual inevitability. But both doctors and patients will be better off to know what's behind the curtain than to blindly accept a new era of algorithmic medicine. You'll be fully armed when you visit with Dr. Algorithm.

chapter four

THE SKINNY ON DEEP LEARNING

The AI revolution is on the scale of the Industrial Revolution—probably larger and definitely faster.

—KAI-FU LEE

AI is probably the most important thing humanity
   has ever worked on.
AI is . . . more profound than electricity or fire.

—SUNDAR PICHAI

IN FEBRUARY 2016, A SMALL START-UP COMPANY CALLED ALIVE-Cor hired Frank Petterson and Simon Prakash, two Googlers with AI expertise, to transform their business of smartphone electro-cardiograms (ECG). The company was struggling. They had developed the first smartphone app capable of single-lead ECG, and, by 2015, they were even able to display the ECG on an Apple Watch. The app had a "wow" factor but otherwise seemed to be of little practical value. The company faced an existential threat, despite extensive venture capital investment from Khosla Ventures and others.

But Petterson, Prakash, and their team of only three other AI talents had an ambitious, twofold mission. One objective was to develop an algorithm that would passively detect a heart-rhythm disorder, the other to determine the level of potassium in the blood, simply from the ECG captured by the watch. It wasn't a crazy idea, given whom AliveCor had just hired. Petterson, AliveCor's VP of engineering, is tall, blue-eyed, dark-haired with frontal balding, and, like most engineers, a bit introverted. At Google, he headed up YouTube Live, Gaming, and led engineering for Hangouts. He previously had won an Academy Award and nine feature film credits for his design and development software for movies including the Transformers, Star Trek, the Harry Potter series, and *Avatar*. Prakash, the VP of products and design, is not as tall as Petterson, without an Academy Award, but is especially handsome, dark-haired, and brown-eyed, looking like he's right out of a Hollywood movie set. His youthful appearance doesn't jibe with a track record of twenty years of experience in product development, which included leading the Google Glass design project. He also worked at Apple for nine years, directly involved in the development of the first iPhone and iPad. That background might, in retrospect, be considered ironic.

Meanwhile, a team of more than twenty engineers and computer scientists at Apple, located just six miles away, had its sights set on diagnosing atrial fibrillation via their watch. They benefited from Apple's seemingly unlimited resources and strong corporate support: the company's chief operating officer, Jeff Williams, responsible for the Apple Watch development and release, had articulated a strong vision for it as an essential medical device of the future. There wasn't any question about the importance and priority of this project when I had the chance to visit Apple as an advisor and review its progress. It seemed their goal would be a shoo-in.

The Apple goal certainly seemed more attainable on the face of it. Determining the level of potassium in the blood might not be something you would expect to be possible with a watch. But the era of deep learning, as we'll review, has upended a lot of expectations.

The idea to do this didn't come from AliveCor. At the Mayo Clinic, Paul Friedman and his colleagues were busy studying details of a part of an ECG known as the T wave and how it correlated with blood levels of potassium. In medicine, we've known for decades that tall T waves could signify high potassium levels and that a potassium level over 5.0 mEq/L is dangerous. People with kidney disease are at risk for developing these levels of potassium. The higher the blood level over 5, the greater the risk of sudden death due to heart arrhythmias, especially for patients with advanced kidney disease or those who undergo hemodialysis. Friedman's findings were based on correlating the ECG and potassium levels in just twelve patients before, during, and after dialysis. They published their findings in an obscure heart electrophysiology journal in 2015; the paper's subtitle was "Proof of Concept for a Novel 'Blood-Less' Blood Test."[1] They reported that with potassium level changes even in the normal range (3.5–5.0), differences as low as 0.2 mEq/L could be machine detected by the ECG, but not by a human-eye review of the tracing.

Friedman and his team were keen to pursue this idea with the new way of obtaining ECGs, via smartphones or smartwatches, and incorporate AI tools. Instead of approaching big companies such as Medtronic or Apple, they chose to approach AliveCor's CEO, Vic Gundotra, in February 2016, just before Petterson and Prakash had joined. Gundotra is another former Google engineer who told me that he had joined AliveCor because he believed there were many signals waiting to be found in an ECG.[2] Eventually, by year's end, the Mayo Clinic and AliveCor ratified an agreement to move forward together.

The Mayo Clinic has a remarkable number of patients, which gave AliveCor a training set of more than 1.3 million twelve-lead ECGs gathered from more than twenty years of patients, along with corresponding blood potassium levels obtained within one to three hours of the ECG, for developing an algorithm. But when these data were analyzed it was a bust (Figure 4.1).

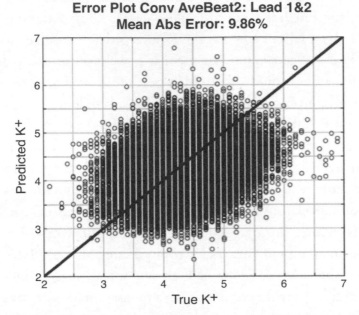

FIGURE 4.1: Plot of data for Mayo Clinic predicted potassium (K+) values from electro-cardiogram versus actual lab-based values. Source: Data from AliveCor.

Here, the "ground truths," the actual potassium (K+) blood levels, are plotted on the x-axis, while the algorithm-predicted values are on the y-axis. They're all over the place. A true K+ value of nearly 7 was predicted to be 4.5; the error rate was unacceptable. The AliveCor team, having made multiple trips to Rochester, Minnesota, to work with the big dataset, many in the dead of winter, sank into what Gundotra called "three months in the valley of despair" as they tried to figure out what had gone wrong.

Petterson and Prakash and their team dissected the data. At first, they thought it was likely a postmortem autopsy, until they had an idea for a potential comeback. The Mayo Clinic had filtered its massive ECG database to provide only outpatients, which skewed the sample to healthier individuals and, as you would expect for people walking around, a fairly limited number with high potassium levels. What if all the patients who were hospitalized at the time were analyzed? Not only would this yield a higher proportion of people with

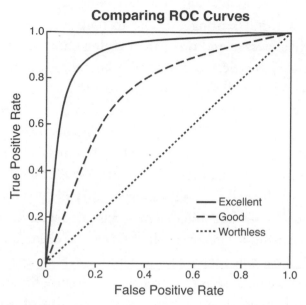

FIGURE 4.2: The receiver operating characteristic (ROC) curves of true versus false positive rates, with examples of worthless, good, and excellent plotted. Source: Adapted from "Receiver Operating Characteristic," *Wikipedia* (2018): https://en.wikipedia.org/wiki/Receiver_operating_characteristic.

high potassium levels, but the blood levels would have been taken closer to the time of the ECG.

They also thought that maybe all the key information was not in the T wave, as Friedman's team had thought. So why not analyze the whole ECG signal and override the human assumption that all the useful information would have been encoded in the T wave? They asked the Mayo Clinic to come up with a better, broader dataset to work with. And Mayo came through. Now their algorithm could be tested with 2.8 million ECGs incorporating the whole ECG pattern instead of just the T wave with 4.28 million potassium levels. And what happened?

Eureka! The error rate dropped to 1 percent, and the receiver operating characteristic (ROC) curve, a measure of predictive accuracy where 1.0 is perfect, rose from 0.63 at the time of the scatterplot (Figure 4.1) to 0.86. We'll be referring to ROC curves a lot throughout the book, since they are considered one of the best ways to show (underscoring *one*, and to point out the method has been sharply criticized and there are ongoing efforts to develop better performance metrics) and quantify accuracy—plotting the true positive rate against the false positive rate (Figure 4.2). The value denoting accuracy is the area under the curve, whereby 1.0 is perfect, 0.50 is the diagonal line "worthless," the equivalent of a coin toss. The area of 0.63 that AliveCor initially obtained is deemed poor. Generally, 0.80–.90 is considered good, 0.70–.80 fair. They further prospectively validated their algorithm in forty dialysis patients with simultaneous ECGs and potassium levels. AliveCor now had the data and algorithm to present to the FDA to get clearance to market the algorithm for detecting high potassium levels on a smartwatch.

There were vital lessons in AliveCor's experience for anyone seeking to apply AI to medicine. When I asked Petterson what he learned, he said, "Don't filter the data too early. . . . I was at Google. Vic was at Google. Simon was at Google. We have learned this lesson before, but sometimes you have to learn the lesson multiple times. Machine learning tends to work best if you give it enough data and the rawest data you can. Because if you have enough of it, then it should be able to filter out the noise by itself."[3]

"In medicine, you tend not to have enough. This is not search queries. There's not a billion of them coming in every minute. . . . When you have a dataset of a million entries in medicine, it's a giant dataset. And so, the order or magnitude that Google works at is not just a thousand times bigger but a million times bigger." Filtering the data so that a person can manually annotate it is a terrible idea. Most AI applications in medicine don't recognize that, but, he told me, "That's kind of a seismic shift that I think needs to come to this industry."[4]

We can see some fundamentals for deep learning algorithmic development. Getting the labeling or ground truths right is critical: the input to the algorithm has to be correct if there's any chance for an algorithm output to be useful. When the potassium values were not taken close to the time of the ECG, the chance of making an accurate prediction was significantly impaired. Filtering the patient sample to only outpatients—because it seemed at first to the people involved to be the best cohort to analyze—nearly killed the project, as did the human assumption that the valuable input would be in the T wave rather than the whole ECG signal.

Deep learning AI is all about inputs and outputs. As Andrew Ng, a rock star in AI, puts it, "Input and output mapping is a new superpower." Algorithms like to eat data, the more the better, but that data needs to include lots of the whole range of values for inputs to get a bull's-eye for outputs. It reminds me of taking care of patients in the intensive care unit: one of the key metrics to assess every day is how much fluid a patient takes in versus how much urine is made. Without accurate records of inputs and outputs, we might treat the patient improperly, perhaps upping the intravenous fluids or prescribing diuretics. In both cases, bad inputs and outputs can kill a patient.

In parallel to the smartwatch potassium project, AliveCor was pursuing detection of atrial fibrillation, the heart rhythm that underlies a risk for stroke. Our lifetime risk of developing atrial fibrillation (AF) is over 30 percent, and the risk of stroke in people who have AF is 3 percent per year. Accordingly, diagnosing a spell of AF in a person without symptoms could be quite important, revealing the potential need for preventive blood thinners. Back in 2015, their cardiologist founder, David Albert, presented the idea at a scientific meeting. The approach was for deep learning about a person's expected heart rate when at rest and during activity to give a range of possible safe heart rates for the individual. Let's say your heart rate while sitting for a while is usually 60, but suddenly it ramps up to 90. If your watch's accelerometer indicates that you're still sitting, the algorithm would kick in for an abnormality, alerting a watch wearer to record an ECG. That would be accomplished by placing a thumb on the watchband.

Prakash, Petterson, and the other three members of the AliveCor AI team developed a deep learning algorithm they called SmartRhythm, a neural network that used a person's most recent five minutes of activity. To be practical, this algorithm has to be integrated with a smartwatch or another device that can assess heart rate on a continuous basis. When Apple released its first watch in 2015, it could only record heart rate for five hours. But the Apple Watch 2 and 3 versions have that ability, using their optical plethysmography sensor (the flashing green lights in back of the watch, used by many manufacturers like Fitbit). The improved battery life on the Apple Watch 3 allows for twenty-four-hour continuous heart rate recording—exactly what SmartRhythm needed to piggyback on. At least during waking hours, there was, for the first time, a way for a person without symptoms to diagnose atrial fibrillation and reduce the risk of stroke.

On November 30, 2017, less than a year and a half after Petterson and Prakash joined the company, the FDA approved Alive-Cor's Kardia band, which substitutes for an Apple Watch band and helps the user detect AF, alerting "possible AF." It was the first FDA-approved AI algorithm to aid consumers in a medical self-diagnosis.

Meanwhile, Apple, aware of the timing of AliveCor's Kardia announcement, announced the same day the launch of a big clinical trial called the Apple Heart Study, conducted in collaboration with Stanford University, to use its heart rate sensor for AF detection.[5] In Apple's case, detection is based on an irregular heart rate, which prompts a telemedicine visit with American Well doctors. Patients are then sent a Band-Aid-like patch to wear that records at least a week of continuous ECG. That's a much more circuitous path for making the AF diagnosis than a thumb on the watchband. The big Apple had just been beaten out by little AliveCor.

At least for about nine months. In September 2018, Apple announced, with much fanfare at its annual hullabaloo, that its AF detection algorithm had been cleared by the FDA for its soon-to-be-released Apple Watch Series 4 and that it was "the first over-the-counter ECG device offered to consumers" and "ultimate guardian for your health." Not exactly, on either count.[6]

The two AliveCor projects for potassium level and AF detection illustrate many of AI's distinctive capabilities, such as detecting things that humans can't, overriding human bias, and providing monitoring on a truly individualized basis. These heart-rhythm and potassium algorithms may seem like small advances, but they represent what can be achieved and translate to some practical value. After all, more than 35 million people wear Apple Watches. Lastly, AliveCor's successes show that, in the era of AI in medicine, David can definitely still beat Goliath.

In this chapter, I want to discuss something of how those advances—and further ones we'll see throughout this book—work. I don't intend to go too deep into deep learning, the technology that underlies a great many technologies. For that, there's an exceptional textbook called *Deep Learning* by Ian Goodfellow, a young, brilliant, staff scientist at Google Brain, and his colleagues.[7] My plan is to avoid the nitty-gritty and instead try to cover only what is most germane to medicine about the technologies we examine. However, some anchoring outside medicine is essential, as AI is far more developed outside of medicine than in it. Were it not for the pioneers and their persistence, we wouldn't be in a position to apply AI in medicine. For that reason, I'll review the major precedents here, as well as terms and a timeline.

In Table 4.1 and Figure 4.3, I've provided a glossary of the key terms that I'll use throughout the book. There's one that I want to specifically highlight, both because it's hugely important and because not everyone agrees on exactly what it means. I've already used the word "algorithm" a bunch of times to describe the AliveCor projects. But what is it? I've always thought, in a reductionist sense, that it meant "if this, then that." But since this book is essentially about algorithmic medicine and its impact, we've got to expand upon that. In his book *The Master Algorithm,* my friend Pedro Domingos, a computer science professor at the University of Washington, defines an algorithm as "a sequence of instructions telling a computer what to do," and he specifies that "every algorithm has an input and an output."[8] This is simple, pretty broad, and would include something as basic as tapping numbers into a calculator. But he goes on: "If every algorithm suddenly stopped working, it would be the end of the world as we know it." Clearly algorithms are much more than just "if this, then that"!

Massimo Mazzotti, a professor at UC Berkeley, further elaborates the meaning of algorithm, capturing a range of current AI capabilities:

> Terse definitions have now disappeared, however. We rarely use the word "algorithm" to refer solely to a set of instructions. Rather, the word now usually signifies a program running on a physical machine—as well as *its effects on other systems*. Algorithms have thus become agents, which is partly why they give rise to so many suggestive metaphors. Algorithms now *do* things. They determine important aspects of our social reality. They generate new forms of subjectivity and new social relationships. They are how a billion-plus people get where they're going. They free us from sorting through multitudes of irrelevant results. They drive cars. They manufacture goods. They decide whether a client is creditworthy. They buy and sell stocks, thus shaping all-powerful financial markets. They can even be creative; indeed, according to engineer and author Christopher Steiner, they have already composed symphonies "as moving as those composed by Beethoven."[9]

In *Homo Deus,* Yuval Noah Harari gives algorithms a remarkable prominence and the broadest possible definition—organisms and the human being—that I've seen:

> Present-day dogma holds that organisms are algorithms, and that algorithms can be represented in mathematical formulas. . . . "Algorithm" is arguably the single most important concept in our world. If we want to understand our life and our future, we should make every effort to understand what an algorithm is, and how algorithms are connected with emotions. . . . [E]motions are bio-

chemical algorithms that are vital for the survival and reproduction of all mammals. . . . 99 percent of our decisions—including the most important life choices concerning spouses, careers and habitats—are made by the highly refined algorithms we call sensations, emotions and desires.[10]

He calls this faith in algorithm's power "dataism," and he takes a bleak view of the future, even going so far as to say, "*Homo sapiens* is an obsolete algorithm."[11]

It just took three sources for us to go all over the map. (I'll stick with just one definition for most of the other terms in Table 4.1.) Collectively, however, I think they do a pretty good job of portraying the breadth, color, and importance of the idea of an algorithm. It's also useful to think of algorithms as existing on a continuum from those that are entirely human guided to those that are entirely machine guided, with deep learning at the far machine end of the scale.[12]

Artificial Intelligence—the science and engineering of creating intelligent machines that have the ability to achieve goals like humans via a constellation of technologies
Neural Network (NN)—software constructions modeled after the way adaptable neurons in the brain were understood to work instead of human guided rigid instructions
Deep Learning—a type of neural network, the subset of machine learning composed of algorithms that permit software to train itself to perform tasks by processing multilayered networks of data
Machine Learning—computers' ability to learn without being explicitly programmed, with more than fifteen different approaches like Random Forest, Bayesian networks, Support Vector machine uses, computer algorithms to learn from examples and experiences (datasets) rather than predefined, hard rules-based methods
Supervised Learning—an optimization, trial-and-error process based on labeled data, algorithm comparing outputs with the correct outputs during training
Unsupervised Learning—the training samples are not labeled; the algorithm just looks for patterns, teaches itself

(continues)

Table 4.1 *(continued)*

Convolutional Neural Network—using the principle of convolution, a mathematical operation that basically takes two functions to produce a third one; instead of feeding in the entire dataset, it is broken into overlapping tiles with small neural networks and max-pooling, used especially for images

Natural-Language Processing—a machine's attempt to "understand" speech or written language like humans

Generative Adversarial Networks—a pair of jointly trained neural networks, one generative and the other discriminative, whereby the former generates fake images and the latter tries to distinguish them from real images

Reinforcement Learning—a type of machine learning that shifts the focus to an abstract goal or decision making, a technology for learning and executing actions in the real world

Recurrent Neural Network—for tasks that involve sequential inputs, like speech or language, this neural network processes an input sequence one element at a time

Backpropagation—an algorithm to indicate how a machine should change its internal parameters that are used to compute the representation in each layer from the representation on the previous layer passing values backward through the network; how the synapses get updated over time; signals are automatically sent back through the network to update and adjust the weighting values

Representation Learning—set of methods that allows a machine with raw data to automatically discover the representations needed for detection or classification

Transfer Learning—the ability of an AI to learn from different tasks and apply its precedent knowledge to a completely new task

General Artificial Intelligence—perform a wide range of tasks, including any human task, without being explicitly programmed

TABLE 4.1: Glossary. Source: *Artificial Intelligence and Life in 2030,* S. Panel, ed. (Stanford, CA: Stanford University, 2016); J. Bar, "Artificial Intelligence: Driving the Next Technology Cycle," in *Next Generation* (Zurich: Julius Baer Group, 2017); Chollet, F., *Deep Learning with Python* (Shelter Island, New York: Manning, 2017); T. L. Fonseca, "What's Happening Inside the Convolutional Neural Network? The Answer Is Convolution," *buZZrobot* (2017); A. Geitgey, "Machine Learning Is Fun! Part 3: Deep Learning and Convolutional Neural Networks," *Medium* (2016); Y. LeCun, Y. Bengio, and G. Hinton, "Deep Learning," *Nature* (2015): 521(7553), 436–444; R. Raicea, "Want to Know How Deep Learning Works? Here's a Quick Guide for Everyone," *Medium* (2017); P. Voosen, "The AI Detectives," *Science* (2017): 357(6346), 22–27.

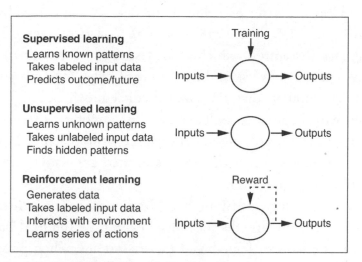

Supervised learning
Learns known patterns
Takes labeled input data
Predicts outcome/future

Unsupervised learning
Learns unknown patterns
Takes unlabeled input data
Finds hidden patterns

Reinforcement learning
Generates data
Takes labeled input data
Interacts with environment
Learns series of actions

Training
Inputs → Outputs

Inputs → Outputs

Reward
Inputs → Outputs

FIGURE 4.3: Schematic explanations of three types of deep learning. Source: Adapted from G. Choy, "Current Applications and Future Impact of Machine Learning in Radiology," *Radiology* (2018): 288(2), 318–328.

A BRIEF HISTORY

With all the chatter and buzz about AI these days, it would be easy to think it was some kind of new invention, but, conceptually, it goes back at least eighty years. In 1936 Alan Turing published a paper on powerful, automated, intelligent systems—a universal computer—titled "On Computable Numbers, with an Application to the *Entscheidungsproblem*."[13] I don't understand the multitude of equations in this thirty-six-page gem, but I must agree with his statement, "We are now in a position to show that the Entscheidungsproblem cannot be solved," both because I can't say it and still don't have a clue what it is! A subsequent paper by Turing in 1950 is viewed as a classic reference for the field of AI.[14]

Several years later, in 1943, Warren McCullogh and Walter Pitts, both electrical engineers, published the first paper describing "logic units" and naming an artificial neuron, the basis and model for what became widely known as neural networks. Given some striking parallels between neurons and electrical circuits, it's no surprise that these pioneering electrical engineers thought of a way to mimic how the brain learns. The term "artificial intelligence" was coined by John McCarthy in 1955. The 1958 *New York Times* coverage of Frank Rosenblatt's Perceptron, which we'd describe in today's terms as a one-layer neural network, is the epitome of hype: "an embryo of an electronic computer" that, the *Times* predicted, "will be able to walk, talk, see, write, reproduce itself and be conscious of its existence." Just a bit later, in 1959, Arthur Samuel used the term "machine learning" for the first time. Other highlights—and there are many—are noted in the timeline (Table 4.2).

Those technologies, despite their prominence today, were not the heart of AI for the first decades of its existence. The field barreled along on the basis of logic-based expert systems, when pessimism took hold among computer scientists, who recognized that the tools weren't working. That souring, along with a serious reduction of research output and grant support, led to the "AI winter," as it became known, which lasted about twenty years. It started to come out of hibernation when the term "deep learning" was coined by Rina Dechter in 1986 and later popularized by Geoffrey Hinton, Yann LeCun and Yoshua Bengio. By the late 1980s, multilayered or deep neural networks (DNN) were gaining considerable interest, and the field came back to life. A seminal *Nature* paper in 1986 by David Rumelhart and Geoffrey Hinton on backpropagation provided an algorithmic method for automatic error correction in neural networks and reignited interest in the field.[15] It turned out this was the heart of deep learning, adjusting the weights of the neurons of prior layers to achieve maximal accuracy for the network output. As Yann LeCun, Hinton's former postdoc, said, "His paper was basically the foundation of the second wave of neural nets."[16] It was Yann a couple of years later who was credited as a father of convolutional neural networks, which are still widely used today for image deep learning.

The public hadn't cued much into the evolving AI story until 1997 when IBM's Deep Blue beat Garry Kasparov in chess. The cover of *Newsweek* called the match "The Brain's Last Stand." Although IBM's name choice of Deep Blue might have suggested it

1936—Turing paper (Alan Turing)

1943—Artificial neural network (Warren McCullogh, Walter Pitts)

1955—Term "artificial intelligence" coined (John McCarthy),

1957—Predicted ten years for AI to beat human at chess (Herbert Simon)

1958—Perceptron (single-layer neural network) (Frank Rosenblatt)

1959—Machine learning described (Arthur Samuel)

1964—ELIZA, the first chatbot

1964—We know more than we can tell (Michael Polany's paradox)

1969—Question AI viability (Marvin Minsky)

1986—Multilayer neural network (NN) (Geoffrey Hinton)

1989—Convolutional NN (Yann LeCun)

1991—Natural-language processing NN (Sepp Hochreiter, Jurgen Schmidhuber)

1997—Deep Blue wins in chess (Garry Kasparov)

2004—Self-driving vehicle, Mojave Desert (DARPA Challenge)

2007—ImageNet launches

2011—IBM vs. *Jeopardy!* champions

2011—Speech recognition NN (Microsoft)

2012—University of Toronto ImageNet classification and cat video recognition (Google Brain, Andrew Ng, Jeff Dean)

2014—DeepFace facial recognition (Facebook)

2015—DeepMind vs. Atari (David Silver, Demis Hassabis)

2015—First AI risk conference (Max Tegmark)

2016—AlphaGo vs. Go (Silver, Demis Hassabis)

2017—AlphaGo Zero vs. Go (Silver, Demis Hassabis)

2017—Libratus vs. poker (Noam Brown, Tuomas Sandholm)

2017—AI Now Institute launched

TABLE 4.2: The AI timeline.

used a DNN algorithm, it was only a rules-based, heuristic algorithm. Nonetheless, this was the first time AI prevailed at a task over a world-champion human, and, unfortunately, framed in this way, it helped propagate the AI-machine versus man war, like the title of the

2017 *New Yorker* piece, "A.I. Versus M.D."[17] The adversarial relationship between humans and their technology, which had a long history dating back to the steam engine and the first Industrial Revolution, had been rekindled.

Kasparov's book, *Deep Thinking*, which came out two decades later, provides remarkable personal insights about that pivotal AI turning point. A month after the match, he had written in *Time* that he thought he could sense "a new kind of intelligence across the table." He recalled, "The scrum of photographers around the table doesn't annoy a computer. There is no looking into your opponent's eyes to read his mood, or seeing if his hand hesitates a little above the clock, indicating a lack of confidence in his choice. As a believer in chess as a form of psychological, not just intellectual, warfare, playing against something with no psyche was troubling from the start." Two of Kasparov's comments about that historic match hit me: One was his observation that "I was not in the mood of playing at all." The other was that "at least [Deep Blue] didn't enjoy beating me."[18] These will be important themes in our discussion of what AI can (and can't) do for medicine.

Even if Deep Blue didn't have much of anything to do with deep learning, the technology's day was coming. The founding of ImageNet by Fei-Fei Li in 2007 had historic significance. That massive database of 15 million labeled images would help catapult DNN into prominence as a tool for computer vision. In parallel, natural-language processing for speech recognition based on DNN at Microsoft and Google was moving into full swing. More squarely in the public eye was man versus machine in 2011, when IBM Watson beat the human *Jeopardy!* champions. Despite the relatively primitive AI that was used, which had nothing to do with deep learning networks and which relied on speedy access to *Wikipedia*'s content, IBM masterfully marketed it as a triumph of AI.

The ensuing decade has seen remarkable machine performance. Deep learning got turbocharged in 2012 with the publication of research by Hinton and his University of Toronto colleagues that

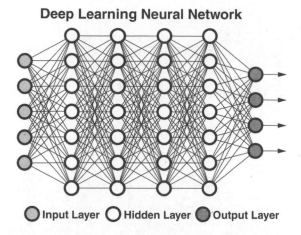

FIGURE 4.4: The architecture of a deep neural network with an input layer, many hidden layers, and the output layer.

showed remarkable progress in image recognition at scale.[19] Progress in unlabeled image recognition was notable in 2012, when Google Brain's team, led by Andrew Ng and Jeff Dean, developed a system based on one hundred computers and 10 million images that could recognize cats in YouTube videos. Facebook's DeepFace was reported to have 97 percent facial recognition accuracy by 2014. For medicine, a landmark *Nature* paper in 2017 on the diagnosis of skin cancer using DNN, matching the accuracy of dermatologists, signified AI's impact on our area of interest.[20] And, as we'll see, despite the misnomers or marketing of Deep Blue and Watson, DNN and related neural networks would come to dominate in games, including Atari, AlphaGo, and poker.

DEEP NEURAL NETWORKS

Much of AI's momentum today—a change as dramatic as evolution's Cambrian explosion 500 million years ago—is tied to the success of deep neural networks. The DNN era, in many ways, would not have come about without a perfect storm of four components. First are the enormous (a.k.a. "big") datasets for training, such as

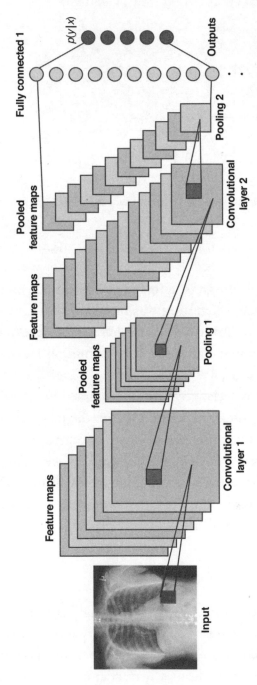

FIGURE 4.5: Schematic of a deep convolutional neural network for chest X-ray interpretation with a series of convolutional layers for feature mapping, pooling, and prediction.

ImageNet's 15 million labeled images; YouTube's vast video library, which grows by three hundred hours of video every minute; Tesla's collection of driving data, which adds 1 million miles of driving data every hour; the airlines' collection of flight data, which grows by 500 Gb with each plane flight; or Facebook's library of billions of images or 4.5 billion language translations a day.[21] Second are the dedicated graphic processing units (GPUs) to run computationally intensive functions with massive parallel architecture, which originated in the video gaming industry. A 2018 publication of optical, diffractive deep neural network (D²NN) prompted Pedro Domingos to say, "Move over GPUs. We can now do deep learning at the speed of light."[22] Third are cloud computing and its ability to store massive data economically. And fourth are the open-source algorithmic development modules like Google's TensorFlow, Microsoft's Cognitive Kit, UC Berkeley's Caffe, Facebook's PyTorch, and Baidu's Paddle that make working with AI accessible.

A deep neural network (Figure 4.4) is structured like a club sandwich turned on its side. But instead of the static BLT, we've got data moving through layers of computations, extracting high-level features from raw sensory data, a veritable sequence of computations. Importantly, the layers are not designed by humans; indeed, they are hidden from the human users, and they are adjusted by techniques like Geoff Hinton's backpropagation as a DNN interacts with the data. We'll use an example of a machine being trained to read chest X-rays. Thousands of chest X-rays, read and labeled with diagnoses by expert radiologists, provide the ground truths for the network to learn from (Figure 4.5). Once trained, the network is ready for an unlabeled chest X-ray to be input. The data go through multiple hidden layers of neurons, from five to one thousand, each responding to different features in the X-ray image, such as shapes or edges. As the image (propagated data) goes onto higher layers, the features and structures get more complex. The deeper the network, by number of layers, the more complexity it can draw out of the input image. At the top layer, the neurons have now fully differentiated the features and are ready for output—predicting what the chest X-ray shows, based on its training.[23] DNNs, with this structural backbone, can functionally be regarded as a general-utility, or a general-purpose, technology, much like the steam engine or electricity.[24] And, much like those technologies, these neural networks can be applied to all sorts of problems. Before medicine, these networks were applied principally to four major areas: games, images, voice and speech, and driverless cars. Each has lessons for us as we explore what deep learning can do for medicine.

GAMES

Even before the historic 1997 Deep Blue–Kasparov chess match, AI had prevailed over human experts in other games including Othello, checkers (which has 500 billion-billion possible positions), and Scrabble.[25] But for all of these, AI used rules-based algorithms, sometimes known as GOFAI, for "good old-fashioned AI." That changed in 2015 when DeepMind beat the classic Atari video game Breakout. The first sentence of the *Nature* paper describing the undertaking is an important foreshadowing for subsequent AI DNN: "We set out to create a single algorithm that would be able to develop a wide range of competencies on a varied range of challenging tasks—a central goal of artificial intelligence that has eluded previous efforts." The algorithm integrated a convolutional neural network with reinforcement learning, maneuvering a paddle to hit a brick on a wall.[26] This qualified as a "holy shit" moment for Max Tegmark, as he recounted in his book *Life 3.0:* "The AI was simply told to maximize the score by outputting, at regular intervals, numbers which we (but not the AI) would recognize as codes for which keys to press." According to DeepMind's leader, Demis Hassabis, the strategy DeepMind learned to play was unknown to any human "until they learned it from the AI they'd built." You could therefore interpret this as AI not only surpassing the video game performance of human professionals, but also of its creators.

Many other video games have been taken on since, including forty-nine different Atari games.[27]

A year later, in 2016, DNN AI began taking on humans directly, when a program called AlphaGo triumphed over Lee Sodol, a world champion at the Chinese game of Go. There was plenty of deep learning prep: training on 30 million board positions that occurred in 160,000 real-life games. According to Edward Lasker, a chess grandmaster, "The rules of Go are so elegant, organic, and rigorously logical that if intelligent life forms exist elsewhere in the Universe, they almost certainly play Go."[28] Its elegance might explain why more than 280 million people watched the tournament live. That number, however, is vastly exceeded by the number of possible Go positions—there are about $2.081681994 \times 10^{170}$, which is two hundred quinquinquagintillion—and vastly more than the number of atoms in the universe, which helps explain why it was a much more interesting challenge than a game like checkers or chess.[29] Go has been played for at least 3,000 years, and the game's experts had predicted in 2015 that it would be at least another decade before AI could win. It took combining DNN (supervised and reinforcement learning) with GOFAI, in the latter case a Monte Carlo tree search.[30] The key winning move (move 37), as it turned out, was viewed as highly creative—despite the fact that a machine made it—and perhaps more importantly, it was made in defiance of human wisdom.[31]

That was certainly a monumental AI achievement for such a remarkably complex and ancient game. But it didn't take long before even that achievement was superseded. In the fall of 2017, AlphaGo Zero, the next iteration of algorithm beyond AlphaGo, took the game world by storm.[32] AlphaGo Zero played millions of games against itself, just starting from random moves. In the *Nature* paper, "Mastering the Game of Go Without Human Knowledge," the researchers concluded that "it is possible [for an algorithm] to train to superhuman level, without human examples or guidance, given no knowledge of the domain beyond basic rules." It was also a stunning example of doing more with less: AlphaGo Zero, in contrast to AlphaGo, had fewer than

5 million training games compared with 30 million, three days of training instead of several months, a single neural network compared with two separate ones, and it performed via a single tensor processing unit (TPU) chip compared with forty-eight TPUs and multiple machines.[33]

If that wasn't enough, just a few months later a preprint was published that this same AlphaGo Zero algorithm, with only basic rules as input and no prior knowledge of chess, played at a champion level after teaching itself for only four hours.[34] This was presumably yet another "holy shit" moment for Tegmark, who tweeted, "In contrast to AlphaGo, the shocking AI news here isn't the ease with which AlphaGo Zero crushed human players, but the ease with which it crushed human AI researchers, who'd spent decades hand-crafting ever better chess software."[35]

AI has also progressed to superhuman performance on a similarly hyper-accelerated course in the game of Texas hold'em, the most popular form of poker. Poker is a different animal in an important way: it is an imperfect information game. In a perfect information game, all the players have the same, identical information, a situation called information symmetry. This is the case for Go, Atari video games, chess, and *Jeopardy!* But, with poker, all players do not have full knowledge of past events. The players are dealt private cards and can bluff. Three papers in *Science* tell the story. The first, in January 2015, by the University of Alberta computer science team, used two regret-minimizing algorithms (what they called CFR+, which stands for counterfactual regret minimization) to "weakly" (their word) solve the game, "proving that the game is a winning game for the dealer."[36] The second paper, in February 2017, also from the University of Alberta and collaborators, was about their so-called DeepStack, which, as its name implies, used a DNN to defeat professional poker players.[37]

That only slight AI advantage didn't last very long. As reported in December 2017 in the third *Science* paper, two computer scientists at Carnegie Mellon published their Libratus algorithm with true superhuman performance against top professionals. Like AlphaGo Zero,

Libratus algorithms are not specific to a particular game but apply to imperfect, hidden information games. In contrast, however, to both the DeepStack poker precedent and AlphaGo Zero, no DNNs were used.[38] What Libratus achieved, being able to infer when the world's best poker players were bluffing and to beat them in such a highly complex game, is no small feat. The extraordinary and rapid-fire success of neural networks in gaming has certainly fueled some of the wild expectations for AI in medicine. But the relative importance of games and people's health couldn't be further apart. It's one thing to have a machine beat humans in a game; it's another to put one's health on the line with machine medicine. One of the reasons I cringe seeing the term "game changer" when it's applied to purported medical progress.

IMAGES

ImageNet exemplified an adage about AI: datasets—not algorithms—might be the key limiting factor of human-level artificial intelligence.[39] When Fei-Fei Li, a computer scientist now at Stanford and half time at Google, started ImageNet in 2007, she bucked the idea that algorithms ideally needed nurturing from Big Data and instead pursued the in-depth annotation of images. She recognized it wasn't about Big Data; it was about carefully, extensively labeled Big Data. A few years ago, she said, "I consider the pixel data in images and video to be the dark matter of the Internet."[40] Many different convolutional DNNs were used to classify the images with annual ImageNet Challenge contests to recognize the best (such as AlexNet, GoogleNet, VGG Net, and ResNet). Figure 4.6 shows the progress in reducing the error rate over several years, with ImageNet wrapping up in 2017, with significantly better than human performance in image recognition. The error rate fell from 30 percent in 2010 to 4 percent in 2016. Li's 2015 TED Talk "How We're Teaching Computers to Understand Pictures" has been viewed more than 2 million times, and it's one of my favorites.[41]

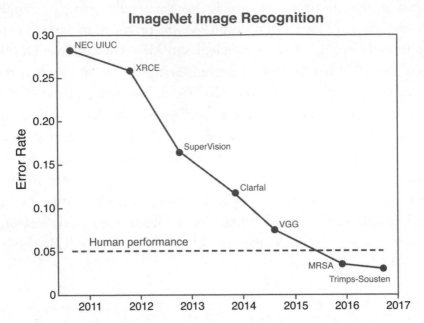

FIGURE 4.6: Over time, deep learning AI has exceeded human performance for image recognition. Source: Adapted from Y. Shoham et al., "Artificial Intelligence Index 2017 Annual Report," *CDN AI Index* (2017): http://cdn.aiindex.org/2017-report.pdf.

The open-source nature of ImageNet's large carefully labeled data was essential for this transformation of machine image interpretation to take hold. Following suit, in 2016, Google made its Open Images database, with 9 million images in 6,000 categories, open source.

Image recognition isn't just a stunt for finding cats in videos. The human face has been at center stage. Just as the accuracy of face recognition soared above 94 percent, so has the controversy attending it, with its potential for invading privacy and fostering discrimination.[42] Apple's Face ID on its iPhone X in 2017 used facial recognition as a biometric password to unlock the phone, as Samsung had previously done. The technology uses the front sensor to scan 3,000 points and make a 3-D model of your face.[43] This has raised privacy concerns. As of 2018, half of American adults have their face images stored in at least one database that police can search, and companies like Karios say they have already read 250 million faces.[44] There are even claims that DNA markers can enable an accurate prediction of a person's face, and therefore identity, leading to sharp rebuke.[45] In contrast, in the opposite direction, facial features can be used to help diagnose rare congenital disease through an AI smartphone Face2Gene app.[46] Other studies suggest an even broader use to facilitate medical diagnoses.[47]

Identifying people using images is not confined to their face. AliveCor developed a four-layered DNN to identify people by their ECG, so that if a user gives a sensor to someone else to use, it'll say, "This doesn't look like you." ECG may even be idiosyncratic enough to serve as a useful biometric, although the dynamic changes that can occur make me question its use for that purpose.

Likewise, there is more to images of the face than just determining personal identity. Back in 2014, UCSD researchers used machine learning of human faces for determining pain, which was shown to be more accurate than human perception.[48] Beyond quantifying pain, there is considerable potential to do the same for stress and mood, as we'll explore in more depth in Chapter 8.

Image segmentation refers to breaking down a digital image into multiple segments, or sets of pixels, which has relied on traditional algorithms and human expert oversight. Deep learning is now having a significant impact on automating this process, improving both its accuracy and clinical workflow.[49]

VOICE, SPEECH, TEXT RECOGNITION, AND TRANSLATION

Processing words is different from processing pixels, since with an image it's all there at once, whereas words, whether speech or text, come in sequence over time. DNNs have transformed the field, which is known as natural-language processing. Notably, the machine accuracy of speech recognition from phone call audio reached parity with humans in 2017 (Figure 4.7).[50] Microsoft has shown that AI is capable of transcribing speech better than professional stenographers. The

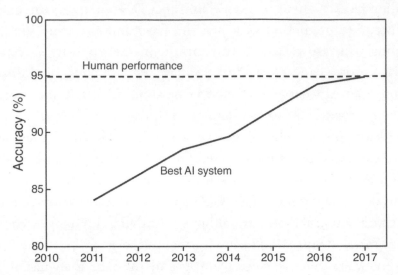

FIGURE 4.7: Over time, deep learning AI has exceeded human performance for voice recognition. Source: Adapted from Y. Shoham et al., "Artificial Intelligence Index 2017 Annual Report," *CDN AI Index* (2017): http://cdn.aiindex.org/2017-report.pdf.

progress has made possible Amazon's Alexa and other voice assistants, which have extensive applications in healthcare. I anticipate that the voice platform will become the basis for a virtual medical coach, and I will lay out the design and features of such a system in Chapter 12.

One of the most striking areas of progress in AI is machine translation. Fernando Pereira, Google's VP and head of Translate, characterized the jump as "something I never thought I'd see in my working life. We'd been making steady progress. This is not steady progress. This is radical."[51] Akin to the AlphaGo Zero algorithm being deployed for many games besides Go, Google, in 2017, published a single translation system capable of transfer learning, a step toward "universal interlingua." By the end of 2016, more than five hundred million monthly users needed 140 billion words per day in a different language.[52] Google translates thirty-seven languages via text and thirty-two via voice, with ever-increasing capabilities to cover more than one hundred languages. Certainly, this and parallel efforts of language translation can be regarded as a pivotal contribution of AI for facilitating human communication.

Machine recognition of text itself, including handwriting, the ability for algorithms to summarize (note I am not using the word "understand") lengthy text, and generating voice from text have all made considerable headway, too.[53] Google's WaveNet and Baidu's Deep Speech are examples of DNNs that automatically generate voice.[54] Noteworthy is the ability to transform text to speech that is indistinguishable from a human voice.[55]

Both facial recognition and other biometrics have interesting applications for cars: they could be used to establish identity in starting the car. Detection of the emotional state or sleepiness level of the driver via voice and facial cues could also be used to promote safety.[56] Still, with more than 40,000 fatal car accidents in the United States in 2017, nearly all due to human error, the biggest payoff of AI in cars might be in driving them for us. Clearly, we're in need of some help.[57]

DRIVERLESS CARS

You'd have to be living in a cave in recent years to avoid being barraged with grandiose claims about driverless cars. Especially if you watch YouTube videos of Tesla "drivers" playing games, writing, jumping into the back seat, and reading, it would be easy to have the sense that they are just around the corner.[58] While it's the pinnacle achievement to date for AI, it's not exactly how it seems.

The Society of Automotive Engineers has a five-level hierarchy of driverlessness, with Level 5 autonomy (Figure 4.8) indicating that a car is fully autonomous—not only does the car do everything, going everywhere, any time, under all conditions, but the human in the car is unable to assume control. That is a long way off, estimated to be decades, if ever attainable.[59] Level 4 means the car is autonomous in most conditions, without the possibility for human backup. The potential for human takeover of the car—conditional automation—is Level 3. Most people are familiar with Level 2, which is like cruise control or lane keeping, representing very limited automation.

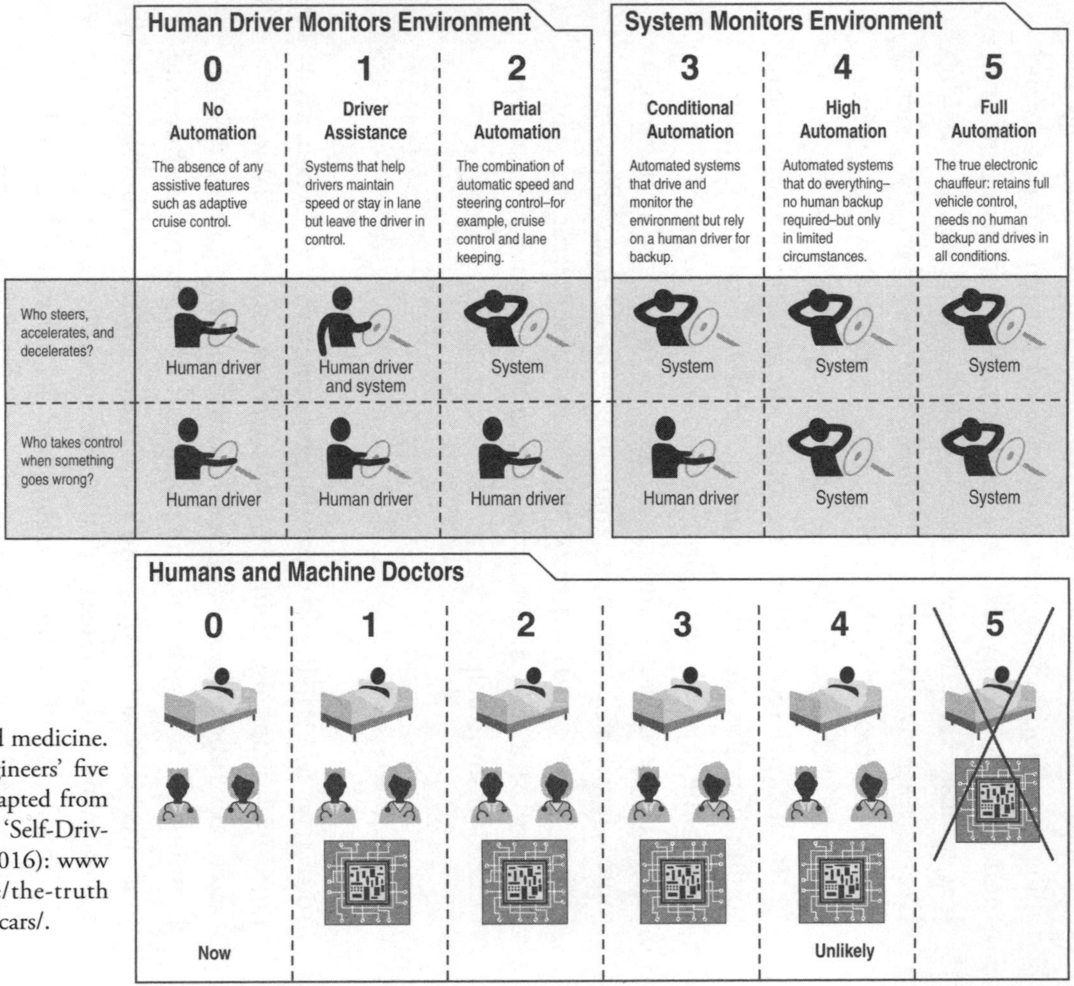

FIGURE 4.8: Self-driving cars and medicine. The Society of Automotive Engineers' five levels of self-driving. Source: Adapted from S. Shladover, "The Truth About 'Self-Driving' Cars," *Scientific American* (2016): www .scientificamerican.com/article/the-truth -about-ldquo-self-driving-rdquo-cars/.

The whole auto industry clearly has its sights on Level 4—with limited need for human backup—which relies on multiple, coordinated technologies. The integrated, multitasking deep learning tracks other cars, pedestrians, and lane markings. Car perception is achieved by a combination of cameras, radar, UDAR (light pulses reflected off objects), and the AI "multi-domain controller" that handles, with DNN, the inputs and the outputs of decisions. Simulating human perceptive capabilities through software is still considered a formidable challenge. Computer vision has since reduced its error rate at identifying a pedestrian from 1 out of 30 frames to 1 in 30 million frames. There's the power of fleet learning to help, whereby the communication and sharing among all autonomous cars with the same operating system can make them smarter. There are other challenges besides perception, however. Even though Level 4 allows for human intervention, cars operating at that level would face catastrophic failure if they experienced the equivalent of a laptop freeze or a web browser crash.

I want to flag this parallel—self-driving cars and the practice of medicine with AI—as one of the most important comparisons in the book. While Level 4 for cars may be achievable under ideal environmental and traffic conditions, it is unlikely that medicine will ever get beyond Level 3 machine autonomy. Certain tasks might be achieved by AI, like accurately diagnosing a skin lesion or ear infection via an algorithm. But, for medicine as a whole, we will never tolerate lack of oversight by human doctors and clinicians across *all* conditions, *all* the time. Level 2—partial automation, like cruise control and lane keeping for drivers—will be of great assistance for both doctors and patients in the future. Having humans serve as backup for algorithmic diagnosis and recommendations for treatment represents conditional automation, and over time this Level 3 autonomy for some people with certain conditions will be achievable.

If these big four AI areas (games, images, speech, cars) summarized here weren't enough, there's a long list of miscellaneous tasks

Beat CAPTCHA	Distinguish fake vs. real art
Create new musical instruments	Autonomous stores
Determine art history	Sort LEGO pieces
Solve Rubik's cube	Make fake videos, photos
Manage stock portfolios	Predict purchase 1 week before person buys it
Write *Wikipedia* articles	
Lip read	Convert text to art
Design websites	Artificial comedy
Tailor clothes	Create slow mode video by imputing frames
Write songs	
Find energy materials	Draw
Brain "shazam" (fMRI music)	Check NDAs
Write text	Pick ripe fruit
Original paintings	Count and identify wild animals
Define accents	Put together IKEA furniture
Write poetry	Create movie trailers
Do the census	Sense human posture through walls
Text to speech w/ accent	
Recommend fashion	Debate
	Predict earthquake aftershocks

TABLE 4.3: Miscellaneous tasks that AI has been reported to achieve in recent years.

that AI has recently been reported to do, some of which are noted in Table 4.3.

If you weren't impressed before, I hope this summary transmits the pluripotent power of AI and gives the historical basis for what has been achieved, especially the progress that has accelerated in recent years. But there are many concerns that give balance to the buzz. In the next chapter, I systematically examine the many liabilities of AI, titrating some of the remarkable exuberance that has characterized the field, and take stock of what it all means for medicine.

chapter five

DEEP LIABILITIES

AIs are nowhere near as smart as a rat.

—Yann LeCun

I often tell my students not to be misled by the name
"artificial intelligence"—there is nothing artificial about it.
AI is made by humans, intended to behave by humans, and,
ultimately, to impact human lives and human society.

—Fei-Fei Li

W HEN I VISITED FEI-FEI LI AT GOOGLE NEAR THE END OF 2017, with AI hype seemingly near its peak, she suggested that we may need another AI winter to get things to cool off, to do reality testing, and to "bubble wrap" the progress that has been made. There's no question that we've seen hyperbole across the board, with forecasts of imminent doom, massive job loss, and replacement of doctors, just to name a few. But when I thought about and researched all the negative issues related to AI, I realized that I could write a whole book on this topic. I'm sure there will be many of those books to come. It should be no surprise to anyone that such a powerful tool as AI could be used for nefarious or adverse purposes, be it maliciously or unwittingly. We'll take an abridged view of the potentially adverse issues confronting deep medicine, ranging from issues with the methodology of AI itself, its potential to enhance bias and inequities, blurring of truth, invasion of privacy, and threats to jobs and even our existence.

NEURAL NETWORK METHODOLOGY AND LIMITS

When I use the term "methodology," I'm referring to everything from input to output, and the output of outputs. Neural networks benefit from optimal quality and quantity of the data that they are trained with and that they use to make predictions. Most AI work to date has been with structured data (such as images, speech, and games) that are highly organized, in a defined format, readily searchable, simple to deal with, store and query, and fully analyzable. Unfortunately, much data is not labeled or annotated, "clean," or structured. In medicine, there's a plethora of unstructured data such as free text in electronic medical records. With rare exceptions, AI, to date, has used supervised learning, which strongly requires establishing "ground truths" for training. Any inaccurate labeling or truths can render the network output nonsensical. For example, doctor interpretation of scans often lacks concordance, making ground truths shaky. Cleaning the data means either getting all the incomplete, irrelevant, corrupt, inaccurate, incorrect stuff out or modifying it to warrant inclusion.

Even when data have been cleaned, annotated, and structured, problems can still crop up. There's the dimension of time to consider: data can drift with models dropping performance as the data change over time.[1]

The AliveCor story illustrates how data selection can introduce bias when, initially, the Mayo Clinic intentionally filtered out all the hospitalized patients. The insufficient number of patients with high potassium levels in the first dataset almost led the team to abandon the project. Overall, there has to be enough data to override noise-to-signal issues, to make accurate predictions, and to avoid overfitting, which is essentially when a neural network comes to mirror a limited dataset. To repeat, there is clearly abundant data when you think of the number of Google searches, Instagram or Facebook posts, or YouTube videos. But for medicine, instead of billions of data points, we're typically in the thousands, occasionally in the millions. Such datasets don't require a deep neural network, and if one is used, there are considerable problems of insufficient input and questionable output.

Although DNNs are routinely positioned as modeling learning capability from our approximately three-pound brain with 86 billion neurons and their 100 trillion interconnections, it needs to be emphasized that there's really nothing to support that assertion. Neural networks aren't actually very neural. François Chollet, a deep learning expert at Google, points out in *Deep Learning with Python,* "There is no evidence that the brain implements anything like the learning mechanisms in use in modern deep learning models."[2] Of course, there's no reason why machines should mimic the brain; that's simplistic reverse-anthropomorphic thinking. And, when we see machines showing some semblance of smartness, we anthropomorphize and think that our brains are just some kind of CPU equivalent, cognitive processing units.

Deep learning AI is remarkably different from and complementary to human learning. Take child development: Yann LeCun, the AI pioneer at Facebook, weighed in on this key issue: "Human children are very quick at learning human dialogue and learning common sense about the world. We think there is something we haven't uncovered yet—some learning paradigm that we haven't figured out. I personally think being able to crack this nut is one of the main obstacles to making real progress in AI."[3]

Machines require big datasets to learn, whereas kids need very little input. It's not just that machines are good at deep learning, and children, using Bayesian probabilistic methods, can do plenty of inferring and extrapolating. Children can reason. They quickly develop an understanding of how the world works. And they show an adaptability to novelty when not having experienced a situation. It takes infants just minutes to learn abstract language-like rules from minimal unlabeled examples.[4] As Microsoft's Harry Shum put it, "Computers today can perform specific tasks very well, but when it comes to general tasks, AI cannot compete with a human child." This brings to mind John Locke's notion of the tabula rasa (which dates back to Aristotle, c. 350 BC) in his classic 1689 *Essay Concerning Human Understanding* (it's four books, not really an essay!), which holds that humans are born with minds like blank slates. Certainly a neural network is like a blank slate, but many researchers, such as Gary Marcus, argue that until artificial intelligence researchers account for human innateness and prewiring, computers will be incapable of feats such as becoming conversant at the same speed as children.[5] So although computers can become unmatched experts at narrow tasks, as Chollet put it, "There's no practical path from superhuman performance in thousands of narrow vertical tasks to the general intelligence and common sense of a toddler." It's the combination of AI learning with key human-specific features like common sense that is alluring for medicine.

All too commonly we ascribe the capability of machines to "read" scans or slides, when they really can't read. Machines' lack of understanding cannot be emphasized enough. Recognition is not understanding; there is zero context, exemplified by Fei-Fei Li's TED Talk on computer vision. A great example is the machine interpretation of "a man riding a horse down the street," which actually is a man on a horse sitting high on a statue going nowhere. That symbolizes the plateau we're at for image recognition. When I asked Fei-Fei Li in 2018 whether anything had changed or improved, she said, "Not at all."

There are even problems with basic object recognition, exemplified by two studies. One, called the "The Elephant in the Room," literally showed the inability for deep learning to accurately recognize the image of an elephant when it was introduced to a living room scene that included a couch, a person, a chair, and books on a shelf.[6]

On the flip side, the vulnerability of deep neural networks was exemplified by seeing a ghost—identifying a person who was not present in the image.[7]

Some experts believe that deep learning has hit its limits and it'll be hard-pressed to go beyond the current level of narrow functionality. Geoffrey Hinton, the father of deep learning, has even called the entire methodology into question.[8] Although he invented backpropagation, the method for error correction in neural networks, he recently said he had become "deeply suspicious" of backprop, saying his view had become that we should "throw it all away and start again."[9] Pointing to the technology's reliance on extensive labeling, he projected that the inefficiencies resulting from that dependence "may lead to their demise."[10] Hinton is intent on narrowing the chasm between AI and children and has introduced the concept of capsule networks.[11] He's clearly excited about the idea of bridging biology and computer science, which for him requires going beyond the flat layers of today's deep neural networks: capsule networks have vertical columns to simulate the brain's neocortex. While capsule architecture has yet to improve network performance, it's helpful to remember that backprop took decades to be accepted. It's much too early to know whether capsule networks will follow suit, but just the fact that he has punched holes in current DNN methodology is disconcerting.

The triumph of AlphaGo Zero also brings up several issues. The *Nature* paper was announced with much fanfare; the authors made the claim in the title "Mastering the Game of Go Without Human Knowledge."[12] When I questioned Gary Marcus on this point, he said that was "ridiculous." "There's a team of seventeen people that are world experts in computer Go. One of them is one of the world's great Go players. Yet, somehow, they said with a straight face 'without human knowledge,' which is absurd." I then asked why he thinks DeepMind hypes it up. To that he said, "They're a very press hungry organization."[13] Marcus isn't alone in the critique of AlphaGo Zero. A sharp critique by Jose Camacho Collados made several key points including the lack of transparency (the code is not publicly available), the overreach of the author's claim of "completely learning from 'self-play,'" considering the requirement for teaching the game rules and for some prior game knowledge, and the "responsibility of researchers in this area to accurately describe . . . our achievements and try not to contribute to the growing (often self-interested) misinformation and mystification of the field."[14] Accordingly, some of AI's biggest achievements to date may have been glorified.

Like all types of science, there are concerns about cherry-picking results or lack of reproducibility. Many research studies test a number of neural networks, but we only read in the publications about those that worked. Or the test sets are very different than the validation sets of data. The lack of open data, such as releasing source code, undermines reproducibility, and that remains an issue for many reports. What also strikes me about the field is its weak publication track record. The papers on AI for games cited in this book are noteworthy, all appearing in top-tier journals like *Nature* and *Science*. But most medically relevant reports show up only on sites such as arXiv as preprints, which are not peer reviewed. The progress that is being made in AI should not bypass the time-accepted validation of the expert peer-review process. Further, the majority of medical studies published to date are retrospective, performed in silico, yet to be prospectively validated in a real-world, clinical setting.

BLACK BOXES

If there's one thing that the human brain and AI certainly share, it's opacity. Much of a neural network's learning ability is poorly understood, and we don't have a way to interrogate an AI system to figure out how it reached its output.

Move 37 in the historic AlphaGo match against Lee Sodol is a case in point: the creators of the algorithm can't explain how it happened. The same phenomenon comes up in medical AI. One example is the capacity for deep learning to match the diagnostic capacities of a team of twenty-one board-certified dermatologists in classifying skin lesions as cancerous or benign. The Stanford computer science creators of that algorithm still don't know exactly what features account for its success.[15] A third example, also in medicine, comes from my colleague Joel Dudley at Mount Sinai's Icahn Institute. Dudley led his team on a project called Deep Patient to see whether the data from electronic medical records could be used to predict the occurrence of seventy-eight diseases. When the neural network was used on more than 700,000 Mount Sinai patients, it was able to predict, using unsupervised learning from raw medical record data (with stacked denoising autoencoders), the probability and timing of onset of schizophrenia (among other conditions), something that is extremely hard for doctors to predict. And Dudley said something that sums up the AI black box problem: "We can build these models, but we don't know how they work."[16]

We already accept black boxes in medicine. For example, electroconvulsive therapy is highly effective for severe depression, but we have no idea how it works. Likewise, there are many drugs that seem to work even though no one can explain how. As patients we willingly accept this human type of black box, so long as we feel better or have good outcomes. Should we do the same for AI algorithms? Pedro Domingos would, telling me that he'd prefer one "that's 99 percent accurate but is a black box" over "one that gives me explanation information but is only 80% accurate."[17]

But that is not the prevailing view. The AI Now Institute, launched in 2017 at New York University, is dedicated to understanding the social implications of AI. The number one recommendation of its AI Now report was that any "high-stakes" matters, such as criminal justice, healthcare, welfare, and education, should not rely on black-box AI.[18] The AI Now report is not alone. In 2018, the European Union General Data Protection Regulation went into effect, requiring companies to give users an explanation for decisions that automated systems make.[19]

That gets to the heart of the problem in medicine. Doctors, hospitals, and health systems would be held accountable for decisions that machines might make, even if the algorithms used were rigorously tested and considered fully validated. The EU's "right to explanation" would, in the case of patients, give them agency to understand critical issues about their health or disease management. Moreover, machines can get sick or be hacked. Just imagine a diabetes algorithm that ingests and processes multilayered data of glucose levels, physical activity, sleep, nutrition and stress levels, and a glitch or a hack in the algorithm develops that recommends the wrong dose of insulin. If a human made this mistake, it could lead to a hypoglycemic coma or death in one patient. If an AI system made the error, it could injure or kill hundreds or even thousands. Any time a machine results in a decision in medicine, it should ideally be clearly defined and explainable. Moreover, extensive simulations are required to probe vulnerabilities of algorithms for hacking or dysfunction. Transparency about the extent of and results from simulation testing will be important, too, for acceptance by the medical community. Yet there are many commercialized medical algorithms already being used in practice today, such as for scan interpretation, for which we lack explanation of how they work. Each scan is supposed to be overread by a radiologist as a checkpoint, providing reassurance. What if a radiologist is rushed, distracted, or complacent and skips that oversight, and an adverse patient outcome results?

There's even an initiative called explainable artificial intelligence that seeks to understand why an algorithm reaches the conclusions that it does. Perhaps unsurprisingly, computer scientists have turned to using neural networks to explain how neural networks work. For example, Deep Dream, a Google project, was essentially a reverse deep learning algorithm. Instead of recognizing images, it generated them to determine the key features.[20] It's a bit funny that AI experts systematically propose using AI to fix all of its liabilities, not unlike the surgeons who say, "When it doubt, cut it out."

There are some examples in medicine of unraveling the algorithmic black box. A 2015 study used machine learning to predict which hospitalized patients with pneumonia were at high risk of serious complications. The algorithm wrongly predicted that asthmatics do better with pneumonia, potentially instructing doctors to send the patients with asthma home.[21] Subsequent efforts to understand the unintelligible aspects of the algorithm led to defining each input variable's effect and led to a fix.[22]

It's fair to predict that there will be many more intense efforts to understand the inner workings of AI neural networks. Even though we are used to accepting trade-offs in medicine for net benefit, weighing the therapeutic efficacy and the risks, a machine black box is not one that most will accept yet as AI becomes an integral part of medicine. Soon enough we'll have randomized trials in medicine that validate strong benefit of an algorithm over standard of care without knowing why. Our tolerance for machines with black boxes will undoubtedly be put to the test.

BIAS AND INEQUITIES

In *Weapons of Math Destruction,* Cathy O'Neil observed that "many of these models encoded human prejudice, misunderstanding, and bias into the software systems that increasingly managed our lives."[23] Bias is embedded in our algorithmic world; it pervasively affects perceptions of gender, race, ethnicity, socioeconomic class, and sexual orientation. The impact can be profound, including who gets a job, or even a job interview, how professionals get ranked, how criminal justice proceeds, or whether a loan will be granted. The problem runs deep, as a few examples illustrate.

In a paper titled "Men Also Like Shopping," a team of researchers assessed two image collections, each containing more than 100,000 complex scene photos with detailed labeling.[24] The images displayed a predictable gender bias: shopping and cooking were linked to women;

sports, coaching, and shooting associated with men. The output distortion is evident in the image of a man in the kitchen cooking, labeled "woman." Worse is the problem that image recognition trained on this basis amplifies the bias. A method for reducing such bias in training was introduced, but it requires the code writer to be looking for the bias and to specify what needs to be corrected. Even if this is done, there's already the embedded problem in the original dataset.[25] Another prominent example of gender bias was uncovered by a Carnegie Mellon study that found that men were far more likely to receive Google ads for high-paying jobs than women.[26]

An AI study of a standard enormous corpus of text from the World Wide Web of about 840 billion words showed extraordinary evidence of both gender and racial bias and other negative attitudes such as mental disease or old people's names.[27] Using the web as the source for this study simply brought to the forefront our historic, culturally entrenched biases. When a Google Photos AI app mistakenly tagged black people as gorillas in 2015, it created quite a flap.[28] *ProPublica*'s exposé called "Machine Bias" provided dramatic evidence for a commonly used commercial algorithm that mistakenly predicted black defendants being at higher risk of committing future crimes. The risk scores for white defendants were automatically skewed to lower risk.[29] There's been bias in algorithms police used against the poor for predicting where crimes will occur[30] and against gays in the infamous "gaydar" study of facial recognition for predicting sexual orientation.[31]

There are unanticipated but important ways to engender bias. Take the development of an app called NamePrism that was supposed to identify and prevent discrimination.[32] The app, developed by Stony Brook University with several major Internet firms, was a machine learning algorithm that inferred ethnicity and nationality from a name, trained from millions of names, with approximately 80 percent accuracy. But when the institutional review board and researchers went forward with the project, they didn't anticipate that the app would be used to promote discrimination. And it was.[33]

The lack of diversity of people who work in the leading tech companies, and at the level of senior management, doesn't help this situation at all. The preponderance of white men in many companies makes it much harder to recognize gender bias against women and requires attention—it won't be fixed by an AI algorithm.

The AI Now Institute has addressed bias, recommending that "rigorous pre-release trials" are necessary for AI systems "to ensure that they will not amplify biases and errors due to any issues with the training data, algorithms, or other elements of system design." And there needs to be continued surveillance for any evidence of bias, which some groups using AI hope to achieve.[34] Kate Crawford, the director, summed it up: "As AI becomes the new infrastructure, flowing invisibly through our daily lives like the water in our faucets, we must understand its short- and long-term effect and know that it is safe for all of us."[35] Efforts have been launched to systematically audit algorithms as a means to promote fairness.[36] Indeed, AI has been used to address *Wikipedia*'s gender bias,[37] and there is even debate as to whether AI algorithms are considerably less biased than people.[38]

Bias in medical research is already baked into the system because patients enrolled in studies are rarely a reflection of the population. Minorities are frequently underrepresented and sometimes not even included in studies at all. In genomic studies, this is especially noteworthy for two reasons: first, people of European ancestry compose most or all of the subjects in large cohort studies, which means that, second, they are of limited value to most people, as so much of genomics of disease and health is ancestry specific. Using such data as inputs into AI algorithms and then applying them for prediction or treatment of all patients would be a recipe for trouble. This has been exemplified by AI of skin cancers, with the research to date performed in very few patients with skin of color.[39]

The looming potential for AI to exacerbate the already substantial (and, in many places like the United States, steadily worsening) economic inequities has medical implications as well. Harari, in *Homo Deus,* projected that "twentieth century medicine aimed to

heal the sick, but twenty-first century medicine is increasingly aim-
ing to upgrade the healthy."[40] These concerns are shared by Kai-Fu
Lee, a highly recognized AI expert based in China, who highlighted
the need "to minimize the AI-fueled gap between the haves and
the have-nots, both within and between nations," and underscored
the importance of considering the social impacts of these systems,
both intended and unintended.[41] It's a triple whammy for the lower
socioeconomic class because the biases in AI frequently adversely
affect them, they are most vulnerable to job loss, and access to AI
medical tools may be much harder to come by. We will need considerable forethought and strategies to make validated, impactful AI
tools available to all to override these concerns.

THE BLURRING OF TRUTH

The world of fake news, fake images, fake speeches, and fake videos
is, in part, a product of AI. We've seen how creators of fake news on
Facebook targeted specific people to sway the 2016 US election and
how various companies' Internet ads use AI to seduce (some will
say addict) people. The problem is going to get worse. We used to
talk about how airbrushing and then Photoshopping images could
be used to shape what we see. That manipulation has been taken to
unprecedented levels of fabrication—not just redrawing images but
rewriting reality—using AI tools.

The start-up Lyrebird can fabricate audio that seems to be au-
thentic from short samples of a person's voice;[42] an AI algorithm
called Houdini can hijack audio files, altering them in such a way
that they sound the same to a human but other AI algorithms (such
as Google Voice) detect drastically different words.[43] Porn videos
have been edited using an algorithm to paste celebrity faces, such
as Gal Gadot, the star of *Wonder Woman,* onto other women's bodies.[44] And researchers at the University of Washington used neural
networks to create video, remarkably difficult to detect as a fake, of
President Obama giving a speech that he never gave.[45]

One form of AI, known as generative adversarial networks (GAN), is being applied to such purposes very frequently. GANs were invented by Ian Goodfellow in 2014, who thought that image generation was lagging behind image recognition. Goodfellow's early creative efforts were soon followed up by a team at Nvidia who built a better, more efficient GAN and were able to produce fake celebrities with unprecedented image quality.[46] Indeed, refinements of GAN have proliferated (CycleGAN, DiscoGAN, StarGAN, and pix2pixHD), a development that will make the ability to distinguish real from fake all the more difficult. There is seemingly no limit for the manipulation of all types of content, further blurring the lines of veracity. This is the last thing we need in the era of truth decay.

PRIVACY AND HACKING

There have been many declarations that we have reached the "end of privacy." The progress in facial recognition accuracy hasn't done anything to undermine that assertion. Facial reading DNN algorithms like Google's FaceNet, Apple's Face ID, and Facebook's DeepFace can readily recognize one face from a million others, and we know that half of US adults have their facial images stored in at least one database that is searchable by police. And AI facial data is only one way to identify a person, complemented by AI of genomic data, which was used in finding the Golden State serial killer. Yaniv Erlich, a genomicist and expert on reidentification, asserts, "In the near future, virtually any European-descent US person will be identifiable" via large-scale consumer genomic databases.[47] To add to that, potent biometrics such as retinal images or electrocardiograms could be similarly used in the future. There's also the Orwellian specter of machine vision AI, with the proliferation of surveillance cameras everywhere, markedly facilitating identification and compromising any sense of privacy.

The story of DeepMind, an AI company, and the Royal Free London National Health Foundation Trust from 2017 illustrates the tension in medical circles.[48] In November 2015, the National Health Service (NHS) entrusted DeepMind Technologies (a subsidiary of Google/Alphabet) to transfer a database of electronic patient records, *with* identifiable data but *without* explicit consent, from NHS systems to the company's own. The data encompassed records for 1.6 million UK citizens going back more than five years. The purpose was to develop a smartphone app called Streams to alert clinicians about kidney injury, which is associated with 40,000 deaths per year in England. Addressing the problem was a laudable goal, but at the time DeepMind had little experience in working in healthcare services. There were also clear concerns for giving the company, a division of Google, the largest advertising company in the world, access to such data, although DeepMind made multiple assurances that the data it received would "never be linked or associated with Google accounts, products, or services."[49] That's a key point. When I visited Verily, Google's spin-off company dedicated to healthcare, its senior executives told me that part of the rationale for the new company was to avoid perceptions of linkage between it and Google.

Whatever the company's assurances, there was no way to track what was being done with the massive NHS patient dataset, which included records of drug overdoses, abortions, mental health treatment, and HIV-positive lab tests, and much more. By late 2017, UK regulators determined that the data had not been provided legally.[50] Responding to the concerns, DeepMind ultimately created a digital ledger system to provide an audit trail of any access to any patient data. Ideally, that would have been incorporated from the project's start to provide an assurance of privacy and security.

In the end the Streams app that DeepMind developed, which it provided for free to the NHS, works extremely well—it markedly reduced the time it took to track down relevant patient information related to kidney dysfunction—and was widely embraced by nurses, doctors, and patient advocacy groups. To quote one nurse, Sarah Stanley, using the app: "We have just triaged the patient in less than 30 seconds. In the past, the process would have taken up to four hours."[51] The head of England's Understanding Patient Data, Nicole Perrin, was quite supportive of the project: "I think it is very important that we don't get so hung up on the concerns and the risks that we miss some of the potential opportunities of having a company with such amazing expertise and resources wanting to be involved in health care."[52] Joe Ledsam of DeepMind AI's team added his perspective: "We should be *more* mindful of the risks and safety of models, not less."

The DeepMind case study brings out so many relevant medical privacy issues of Big Data: not obtaining proper consent, not being transparent, and the "techlash" that we'll see more and more with the oligopoly of big tech titans (Google, Amazon, Apple, Facebook, Microsoft) now all fully committed to healthcare. Even though it resulted in an important product that helped clinicians and patients, there were valuable lessons learned.[53]

One other example of deep learning's potential to invade privacy is an effort described in a paper in the *Proceedings of the National Academy of Sciences*.[54] By combining 50 million Google Street View images of 22 million cars, in two hundred cities, with publicly available data, researchers at Stanford University's AI lab and their collaborators were able to accurately estimate public voting patterns, race, education, and income by zip code or precinct. While the use of deep learning algorithms did not provide estimates at the individual or household level, you can be certain that many tech companies have such data and, with similar neural net analytics, can come up with this information. The most notable case is Cambridge Analytica's expansive individual profiles of the majority of American adults, developed via extracting Facebook's data, ultimately with claims of shifting the 2016 election outcome, alongside algorithmically targeted fake news distribution.[55]

The worry about potential hacking has been reinforced by automated cyberattacks and the legitimate concern that products of AI, like driverless cars, could be run off the road. We've already seen how today's nonautonomous cars can be remotely hacked and made to go haywire while being driven.[56] In the era of hacking, all operations using AI need to be concerned about bad data corrupting their system, AI malware, AI bots, and the war of AI versus AI (host system rejecting the invader).

On the flip side, efforts to use DNNs to promote cybersecurity are also underway. Clearly, they haven't taken hold yet, with the massive data breaches at Equifax, Yahoo, Under Armour (MyFitnessPal app), and many others. Perhaps more encouraging is a concept known as differential privacy, which uses a family of machine learning algorithms called Private Aggregation of Teacher Ensembles to preserve the identity of each individual by not ingesting the specific medical history.[57] Yet this use of limited data could bias models to certain subgroups, highlighting the interactions of privacy and bias.

ETHICS AND PUBLIC POLICY

With the way AI has been accelerating in recent years, it should not be surprising that some have called for a speedometer and new regulatory efforts.[58] Oren Etzioni, the CEO for the Allen Institute for AI, for one has called for "steps to at least slow down progress on AI, in the interest of caution." AI didn't create many of the issues discussed in this chapter—they are classic ethical issues—but they are now being amplified, by AI efforts, as illustrated by the "gaydar" study, the NHS-DeepMind collaboration, racial bias, or the unintended promotion of inequities. AI's ethical responses are not necessarily classic, however. There are two fundamental levels of the ethics of AI: machine ethics, which refers to the AI systems per se, and the wider domain, not specific to the algorithms.

The prototypical example of machine ethics involves how driverless cars handle the dilemma of choosing between evils in the case of an impending accident, when no matter how it responds, people are going to die. It's the modern-day version of the trolley problem introduced more than fifty years ago. Jean-Francois Bonnefon and colleagues examined the driverless car dilemma in depth using simulations and input from more than 1,900 people.[59] In each of the three scenarios (Figure 5.1), there is no good choice; it's just a matter of who and how many people are killed, whether the car's passenger, a pedestrian, or several of them. There is no correct answer, with the conflicts of moral values, cultural norms, and personal self-interest, but the majority of respondents did not go for the "greater good" choice of sacrificing themselves. Clearly, trying to deal with these issues in the design of an algorithm to control an autonomous vehicle will be formidable[60] and has been labeled as one of "the thorniest challenges in artificial intelligence today."[61] Another layer of this dilemma is who should be involved in algorithmic design—consumers, manufacturers, government? As you might anticipate, companies are not enthusiastic about government regulation; many firms, including Microsoft and Google, have set up their own internal ethics boards, arguing that regulatory involvement might be counterproductive, delaying the adoption of self-driving cars over fringe issues when it already seems clear that autonomous vehicles will reduce traffic fatalities overall. But we don't think of it in the big picture way. More than 1.25 million people are killed by human drivers each year, most by human error, but we as a society don't bat an eye at the situation.[62] The introduction of computers into the mix sets up a cognitive bias, not acknowledging the net benefit. When a self-driving car kills a person, there's an outcry over the dangers of self-driving cars. The first fatality of a pedestrian hit by a driverless car occurred in an Uber program in Arizona in 2018. The car's algorithm detected a pedestrian crossing the road in the dark but did not stop, and the human backup driver did not react because she trusted the car too much.[63] Here, ironically, I would question the ethics of the company, rather than AI per se, for prematurely pushing the program forward without sufficient testing and human driver backup.

Regulatory AI issues are especially pertinent in medicine. We are in the early days of regulatory oversight of medical algorithms, and only a limited number have been approved. But the issue isn't

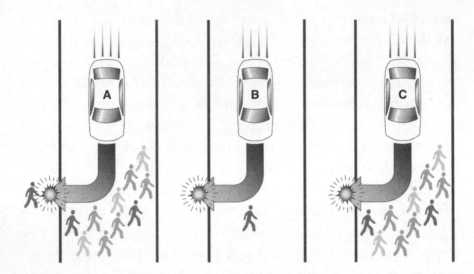

FIGURE 5.1: The three self-driving car traffic situations that result in imminent, unavoidable harm. The car must decide whether to (A) kill several pedestrians or one passerby, (B) kill one pedestrian or its own passenger, or (C) kill several pedestrians or its own passenger. Source: Adapted from J. F. Bonnefon et al., "The Social Dilemma of Autonomous Vehicles," *Science* (2016): 352(6293), 1573–1576.

simply how many AI-based applications have been submitted to the FDA or that a great many more will be submitted for approval. These tools are and will be constantly evolving with larger datasets and autodidactic potential. This will require developing new ground rules for review and approval, conducting post-market surveillance, and bringing on new personnel with AI expertise at regulatory agencies. Giving a green light to an algorithm that is not properly validated, or that is easily hacked, could have disastrous implications.

The concern for ethical breaches and harm led not only to the formation of the AI Now Institute but to many other efforts across the globe to promote the ethics and safety of AI, including OpenAI, Pervade, Partnership on AI, the Future of Life Institute, the AI for Good Summit, and academic efforts at UC Berkeley, Harvard, the University of Oxford, and Cambridge. Yet, as the AI Now Institute has pointed out, there is no tech company tracking its own adherence to ethical guidelines. That hit home for me when I read a recent Infosys AI healthcare report, "AI for Healthcare: Balancing Efficacy and Ethics."[64] Although the report claimed that the industry as a whole and the organizations in it need "to establish ethical standards and obligations," it provided no indication of what those standards or obligations were. In healthcare, there's even the potential to deliberately build algorithms that are unethical, such as basing prediction of patient care recommendations on insurance or income status. It's abundantly clear that there's much more work to be done.

JOBS

I've lost count of the number of articles I've seen entitled "Will AI [or Robots] Take Your Job?" For all these negative prognostications, there seems to be an equal number going in the opposite direction. A go-to source for me is Erik Brynjolfsson, who directs MIT's Initiative on the Digital Economy, and who has said, "Millions of jobs will be eliminated, millions of new jobs will be created and needed, and far more jobs will be transformed."[65] In terms of numbers, Cognizant Technology Solutions projects twenty-one new job categories, 19 million jobs lost, and 21 million new ones created in the next fifteen years.[66] In the same vein, Jerry Kaplan, a Stanford adjunct faculty member who teaches AI, argues that "artificial intelligence will change the way that we live and work, improving our standard of living while shuffling jobs from one category to another in the familiar capitalist cycle of creation and destruction."[67] A 160-page report by the McKinsey Global Institute, agreeing with the overall trade-off, provided an in-depth global perspective, highlighting that the specific jobs gained and lost will vary markedly throughout different regions of the world.[68] It's clear that there will be a shift in and disruption of jobs, and the response to these changes is not as simple as training coal miners to become data miners. Elisabeth Mason, who directs Stanford University's Poverty and Technology Center, thinks that there are millions of unfilled jobs in the United States and that we now have the tools to promote matching—to use AI to help solve this problem.[69] A 2018 Organisation for Economic Co-operation and Development (OECD) report estimates that more than 40 percent of all health-care jobs can be automated across the globe, which underscores the magnitude of disruption we're potentially facing.[70] Within AI, there is a great mismatch between the human talent available and the demand for it. There have been numerous reports of starting salaries for fresh-out-of-school PhDs with AI expertise ranging from $300,000 to more than $1 million; most of these new graduates come from academia or are pinched from other tech companies. There are even jokes that the field needs a National Football League salary cap equivalent for AI specialists.[71] Ultimately, a bigger challenge than finding new jobs for displaced workers (or applicants for jobs in AI) today may be creating new jobs that are not better or, largely, performed by a machine.

We have had and will continue to have a difficult time adjusting to such changes. Garry Kasparov reminds us about the cycle of automation, fear, and eventual acceptance in *Deep Thinking*, describing how, although "the technology for automatic elevators had existed since 1900," it took until the 1950s (after the elevator operators' union strike in 1945) to be accepted because "people were too uncomfortable to ride in one without an operator." Some leaders in the tech industry are stepping up financially to smooth the path toward acceptance. Google, for example, is giving $1 billion to nonprofits that aim to help workers adjust to a new economy.[72] In the chapters that follow I address the job changes, new and old, and transformed, across virtually every type of healthcare worker.

EXISTENTIAL THREAT

We won't have to worry too much about human health and AI if we're no longer here. Whether and when we will ever build autonomous agents with superintelligence that operate like sentient life, designing and building new iterations of themselves, that can accomplish any goal at least as well as humans is unclear. We have clearly been inoculated with the idea, however, after exposure to cumulative doses of Skynet in *The Terminator,* HAL 9000 in *2001: A Space Odyssey,* and Agent Smith in *The Matrix.*

These extremely popular films portrayed sentient machines with artificial general intelligence, and many sci-fi movies have proven to be prescient, so fears about AI shouldn't come as much of a surprise.[73] We've heard doom projections from high-profile figures like Stephen Hawking ("the development of full AI could spell the end of the human race"), Elon Musk ("with AI we are summoning the demon"), Henry Kissinger ("could cause a rupture in history and unravel the way civilization works"), Bill Gates ("potentially more dangerous than a nuclear catastrophe"), and others. Many experts take the opposite point of view, including Alan Bundy of the University of Edinburgh[74] or Yann LeCun ("there would be no *Ex Machina* or *Terminator* scenarios, because robots would not be built with human drives—hunger, power, reproduction, self-preservation").[75] Perhaps unsurprisingly, then, LeCun's employer, Mark Zuckerberg, isn't worried either, writing on Facebook, "Some people fear-monger about how A.I. is a huge danger, but that seems far-fetched to me and much less likely than disasters due to widespread disease, violence, etc."[76] Some AI experts have even dramatically changed their views, like Stuart Russell of UC Berkeley.[77] There's no shortage of futurologists weighing in, one way or another, or even both ways, and even taking each other on.[78] I especially got a kick out of the AI and Mars connection, setting up disparate views between Andrew Ng and Elon Musk. Ng said, "Fearing a rise of killer robots is like worrying about overpopulation on Mars before we populate it,"[79] whereas Musk has said that the potential rise of killer robots was one reason we needed to colonize Mars—so that we'll have a bolt-hole if AI goes rogue and turns on humanity.[80]

Musk's deep concerns prompted him and Sam Altman to found a billion-dollar nonprofit institute called OpenAI with the aim of working for safer AI. In addition, he gave $10 million to the Future of Life Institute, in part to construct worst-case scenarios so that they can be anticipated and avoided.[81] Max Tegmark, the MIT physicist who directs that institute, convened an international group of AI experts to forecast when we might see artificial general intelligence. The consensus, albeit with a fair amount of variability, was by the year 2055. Similarly, a report by researchers at the Future of Humanity Institute at Oxford and Yale Universities from a large survey of machine learning experts concluded, "There is a 50 percent chance of AI outperforming humans in all tasks in 45 years and automating all human jobs in 120 years."[82] Of interest, the Future of Humanity Institute's director Nick Bostrom is the author of *Superintelligence* and the subject of an in-depth profile in the *New Yorker* as the proponent of AI as the "Doomsday Invention."[83] Tegmark points to the low probability of that occurring: "Superintelligence arguably falls into the same category as a massive asteroid strike such as the one that wiped out the dinosaurs."[84]

Regardless of what the future holds, today's AI is narrow. Although one can imagine an artificial general intelligence that will treat humans as pets or kill us all, it's a reach to claim that the moment is upon us: we're Life 2.0 now, as Tegmark classifies us, such that we, as humans, can redesign our software, learning complex new skills but quite limited with respect to modulating our biological hardware. Whether Life 3.0 will come along, with beings that can design both their hardware and software, remains to be seen. For the near future, then, the issue isn't whether human life will become extinct, but how it will change us and our livelihoods—and, for our purposes, how it will change those of us who practice medicine. Tegmark suggests that collectively we begin to think of ourselves as *Homo sentiens*. Let's ask what *Homo sentiens* will look like in a doctor's coat.

chapter six

DOCTORS AND PATTERNS

> If a physician can be replaced by a computer, then he or she deserves to be replaced by a computer.
>
> —WARNER SLACK, HARVARD MEDICAL SCHOOL

> It's sort of criminal to have radiologists doing this, because in the process they are killing people.
>
> —VINOD KHOSLA

SEVERAL YEARS AGO, AFTER A HORRIFIC BOUT OF ABDOMINAL and back pain, I was diagnosed with stones in my left kidney and ureter. Two stones were in the ureter, and they were pretty large—more than 9 millimeters. I was unable to pass them, despite drinking gallons of water and taking drugs like Flomax to increase the chances. The next step was a lithotripsy—a procedure that applies high-energy sound waves from a lithotripter machine outside the body to the ureter region to break up the stones. The procedure is quite painful, although why it is so is poorly understood, and often general anesthesia is used, as it was in my case.

A week after getting the shock waves, I was hopeful for evidence that the stones had broken up or had even disappeared. A simple KUB (kidney, ureter, bladder) X-ray took just a couple of minutes.

The technician who took the picture reviewed it to be sure it was of adequate quality. I asked her whether she could see the stones in my left ureter, and she wasn't sure. In looking at the film myself, I wasn't sure either. So, I asked to speak to the radiologist.

That doesn't happen very often. Unless the radiologist is "interventional," which means a specialist who does treatment procedures (like placing a device in a blood vessel), radiologists rarely have contact with patients, whether to discuss the results of a scan or anything else. The normal operating procedure is that the radiologist, who sits all day in a dark room looking at films, makes the interpretation and files a report. Most physicians are too busy to look at the X-rays they order, but just the report, to review the results with a patient.

It seems a bit like going to buy a car. You talk to the salesperson but never meet the dealer. When you are negotiating with the salesperson, he or she goes to another room to check with the manager. If you've had medical scans done, you've likely never met the reader of the scan, just the photographer. Oftentimes, however, the photographer meets with the reader—either personally or digitally—while you are sitting there to make sure that the scan is technically adequate before you can get dressed and move on.

After sitting in the waiting area for fifteen minutes, I was taken back to the dark room to see the radiologist. Via the reflection of the screen I saw a dark-haired, bearded, prophet-looking white guy, about my age, who was affable and seemed quite pleased to have me visit. He pulled over a chair to have me sit next to him. I was wondering why he was wearing a long white coat and thought he might as well be in jeans and casual clothes. Or even pajamas.

The radiologist digitally pulled up my many kidney-related scans at once on a large screen to do side-by-side comparisons. The CT scan that had led to the original diagnosis a few months back showed the two stones really well, so that scan was the comparison for the new KUB study. The radiologist magnified the new film manyfold to zoom in on the stones. To my chagrin, not only were the stones still there, but they had not changed in size. Only one had migrated downstream a bit. The treatment had resulted in a trivial change. My kidney was still dilated, so there was some obstruction, which raised our concern about how long that could go on without some permanent damage. Despite all the bad news, I got a lot of insights into my condition from this session, much more than I would have gotten from a review session with my urologist. From the radiologist's perspective the lithotripsy procedure was likely a failure, indicating to him that the stones, given their size and location, would almost certainly require a surgical procedure for removal. Plus, because the radiologist had no interest in doing operations, getting the word that it was necessary from him represented a much more independent assessment of my condition, compared to the advice I would have gotten from a surgeon.

This meeting between a patient and a radiologist was an anomaly, but it may be an important indicator of the future.

The past fifty years have introduced important changes to radiology. As the medium moved from analog to digital, it became much easier to store and find films. The images are of much greater resolution. And the speed of digital acquisition of the picture means there's no need to wait for an X-ray film to be developed. It's funny to think back to the clunky, old alternator readers, which were machines that had the hard copy films on them, often filed in alphabetical order of the patients studied that day, among which you waited several minutes for the scan of interest to eventually show up. Attending physicians, along with their trainees on hospital teams, would go to the basement each day to review patient studies directly with the radiologist. But now, with picture archiving and communication systems (PACS), those visits are rare. The referral doctor looks at the report and ideally (although often not) the scan itself, remotely. My friend Saurabh Jha, a radiologist at the University of Pennsylvania, summed up the change well: "The radiologist with a flat plate on the alternator was once the center of the universe. With PACS and advanced imaging, the radiologist is now one of Jupiter's moons, innominate and abundant."[1] The whole process, extending from plain

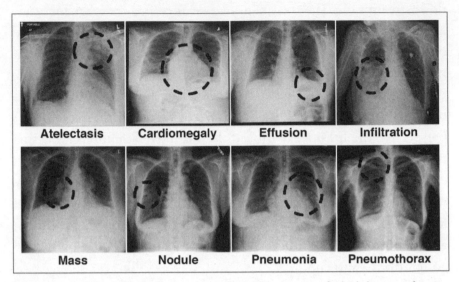

FIGURE 6.1: Eight different findings on a chest X-ray, many of which have overlapping features and can be difficult to use for diagnosis. Source: Adapted from X. Wang et al., *ChestX-ray8: Hospital-Scale Chest X-ray Database and Benchmarks on Weakly-Supervised Classification and Localization of Common Thorax Diseases,* arXiv (2017): https://arxiv.org /abs/1705.02315.

films to CT, PET, nuclear, and MRI scans, has been made more efficient. Except the interpretation.

The classic medical image is the chest X-ray. Two billion are done each year worldwide. They are often tricky to read, especially when used to diagnose pneumonia. Heart failure and many other overlapping features such as scarring, a mass or nodule, fluid, or collapsed lung tissue can confound the diagnosis (Figure 6.1).[2]

Surely a machine that could accurately and quickly read chest X-rays would be a major step forward for the field. As Gideon Lewis-Kraus described it, "A network built to recognize a cat can be turned around and trained on CT scans—and on infinitely more examples than even the best doctor could ever review."[3] Ingesting tons of images is only a small part of the story, of course; the machine needs to learn to interpret them. Despite that looming issue, Geoffrey Hinton proclaimed, "I think that if you work as a radiologist, you are like Wile E. Coyote in the cartoon. You're already over the edge of the cliff, but you haven't yet looked down. There's no ground underneath. People should stop training radiologists now. It's just completely obvious that in five years deep learning is going to do better than radiologists."[4]

Late in 2017, the Stanford computer science group led by Andrew Ng claimed that it had. Ng tweeted, "Should radiologists be worried about their jobs? Breaking news: We can now diagnose pneumonia from chest X-rays better than radiologists." Using a convolutional neural network with 121 layers, learning from more than 112,000 films imaged from over 30,000 patients, the team concluded its algorithm "exceeds average radiologist performance on pneumonia detection."[5] But, notably, the comparison was for only four radiologists, and there were serious methodologic issues.[6] Lior Patcher, a computational biologist at Caltech wrote, "When high profile machine learning people oversell their results to the public it leaves everyone else worse off. And how can the public trust scientists if time and time again they are presented with hype instead of science?"[7]

We surely can't conclude from the data so far that radiologists are an endangered species, as Hinton and Ng, two of the AI world's leaders, suggest. Such assertions have unfortunately characterized many reports on AI in medicine, even though each has the problem of being a retrospective, in silico, non-replicated study, no less with flaws in interpretation of data at many levels. As Declan O'Regan, a radiologist who works on machine learning, wrote to me on Twitter, "Any PhD can train a deep learning network to classify images with apparent human-level performance on cross-validation. Take it into the real world and they will all underperform."

A radiologist reads about 20,000 studies a year, which equates to somewhere between 50–100 per day, a number that has been steadily increasing.[8] While X-rays are single-digit images per exam, ultrasounds are dozens, and CT scans and MRIs are hundreds, a ratio that keeps increasing. All told, there are more than 800 million medical scans a year in the United States, which amounts to about 60 billion images, or one image generated every two seconds.[9] Nevertheless, radiologists, through training and experience, develop pattern recognition visual systems that allow them to rapidly identify abnormalities. An attention researcher at Harvard, Dr. Trafton Drew, said, "If you watch radiologists do what they do,[10] [you're] absolutely convinced that they are like superhuman."[11] It's like System 1 thinking because it is reflexive, pattern-matching rather than logical analysis. But radiologists still suffer from "inattentional blindness"—that is, they can become so focused on looking for specific things that they will miss unexpected data that is literally in front of their noses. This was demonstrated by superimposing a picture of a man in a gorilla suit shaking his fist into images that a bunch of radiologists were reviewing for signs of cancer. It turned out that 83 percent of the radiologists missed the man in the gorilla suit.[12]

Some studies suggest that errors in interpretation of medical scans are far worse than generally accepted, with false positive rates of 2 percent and false negative rates over 25 percent. Given those 800 million annual scans, that means a large number of readings are at risk of being wrong. Notably, 31 percent of American radiologists have experienced a malpractice claim, most of which were related to missed diagnoses.[13]

Radiologists, then, would certainly benefit from a machine accuracy booster. For example, a careful study of classifying more than 50,000 chest X-rays as simply either normal or abnormal achieved algorithmic accuracy of 95 percent, which could prove useful for radiologists to triage which ones merit a closer look.[14] It's not just about lapses of human attention or errors. Time, too, is a major factor: instead of a yearly 20,000 films a radiologist can review, millions to even billions of images could be investigated. For example, when Merge Healthcare, a medical-imaging firm, was acquired by IBM in 2015, its algorithms got access to more than 30 billion images.[15] What's more, there is plenty of information on a medical image—in every pixel or voxel, a pixel's 3-D equivalent—that may not be seen by the human eye, such as the texture, the degree of dye-enhancement, or the intensity of the signal. A whole field, sometimes called radiomics, has developed to investigate the signatures lurking in the scan data.[16] For example, it has led to the development of a metric known as the Hounsfield units of stone density, which reveals the mineral constituent of the stone, such as calcium oxalate or uric acid, as well as indicating the therapy that would most likely be successful. That's perfect for machine reading: algorithms enable a deeper quantification of the data in scans, releasing heretofore unrecognized value.

Several examples bring this pivotal point home. A Mayo Clinic team showed that the texture of brain MRI images could predict a particular genomic anomaly, specifically 1p/19q co-deletion, that's relevant to surviving certain types of brain cancer.[17] Similarly, using deep learning algorithms to read MRI scans of patients with colon cancer could reveal whether a patient has a critical tumor-gene mutation, known as KRAS, awareness of which should significantly influence treatment decisions.[18] Machine learning of mammography images from more than 1,000 patients, coupled with biopsy results indicating a high risk of cancer, showed that more than 30 percent of breast surgeries could be avoided.[19] Applying deep learning to X-ray images of hip fractures can lead to diagnoses as accurate as those derived from the more advanced—and so more expensive—image techniques, including MRI, nuclear bone scans, or CT, which doctors otherwise turn to when analyses of X-rays give uncertain results. Using a convolutional neural network with 172 layers, trained with over 6,000 X-rays (with a total of 1,434,176 parameters), and validated in more than a thousand patients, the accuracy of the algorithm was shown to be greater than 99 percent, quite comparable to performance by experienced radiologists.[20] Multiple reports from academic medical centers have shown the power of deep learning to sort through a variety of scans, including CT scans for liver and lung nodules and bone age, adding to the expanding evidence that machines can accomplish accurate diagnostic work. UCSF developed a 3-D convolutional neural network for chest CT in more than 1,600 patients, of whom 320 had confirmed lung cancer.[21] The University of Tokyo developed a six-layer convolutional neural network for CT liver mass classification from 460 patients with an overall 84 percent accuracy compared with ground truths.[22] Geisinger Health in Pennsylvania used nearly 40,000 head CT scans to show high accuracy of machine diagnosis of brain hemorrhage.[23] Radboud University in the Netherlands found that a deep neural network trained on more than 1,400 digital mammograms gave similarly accurate readings as those performed by twenty-three radiologists.[24] And Stanford University's convolutional neural network used more than 14,000 X-rays to learn how to quantify bone age, giving results as good as those from three expert radiologists.[25] Computer scientists in South Korea at Seoul National University developed and validated a deep learning algorithm using more than 43,000 chest X-rays, trained to detect cancerous lung nodules. The algorithm was remarkably accurate in four retrospective cohorts (AUC = 0.92–0.96). It compared quite favorably to board-certified radiologists and provided additive value as a "second reader" for even higher accuracy when the two were combined.[26]

Surely you don't need a convolutional neural network to get the picture that algorithmic image processing is making great strides.

Academic medical centers are not the only groups pursuing the technology, of course. Deep learning of medical images has been undertaken by many companies including Enlitic, Merge Healthcare, Zebra Medical Vision, Aidoc, Viz.ai, Bay Labs, Arterys, RAD-Logic, Deep Radiology, and Imagen. Each of these companies has made progress in specific types of images. Arterys specializes in heart MRI and received the first FDA approval for AI medical imaging in 2017. In 2018, Viz.ai received FDA approval for deep learning of head CT scans for diagnosis of stroke, with immediate text notification of clinicians. Imagen followed soon thereafter with FDA approval for its machine processing of bone films. For Enlitic, the autodidactic processing of thousands of musculoskeletal scans enabled the company's algorithms not just to make diagnoses of bone fractures at remarkable accuracy, but it was even able to spotlight the sites of micro-fractures when the fracture was as small as 0.01 percent of the X-ray image in question. Zebra Medical Vision validated a convolutional neural network that detects vertebral compression fractures with 93 percent accuracy, whereas radiologists miss such fractures more than 10 percent of the time.[27] This same company used deep learning for heart calcium score prediction.[28] All of these radiology AI companies are marching along to commercialize their algorithmic scan reading capabilities. By late 2017, Zebra Medical Vision was deployed in fifty hospitals and had analyzed more than 1 million scans, at a speed almost 10,000 times greater than can be achieved by radiologists, at a cost of only one dollar per scan.[29]

It's clear that deep learning and machines will have an important role in the future of radiology. Some pronouncements could be considered exuberant, however, such as Andrew Ng's suggestion that radiologists might be easier to replace than their executive assistants,[30] or the conclusion of the essay by Katie Chockley and Ezekiel Emanuel, titled "The End of Radiology?," that radiology may disappear as a thriving specialty in the next five to ten years.[31] The venture capitalist Vinod Khosla has said, "The role of the radiologist will be obsolete in five years." I know Vinod well and have discussed this issue with him. He didn't mean radiologists would be obsolete but that their current role as primary readers of scans would. At the other extreme, Emanuel, the prominent physician and architect of the Affordable Care Act, in a *Wall Street Journal* article, claimed that "machine learning will replace radiologists and pathologists, interpreting billions of digital X-rays, CT and MRI scans and identifying abnormalities in pathology slides more reliably than humans."[32]

Even if it is easy to get carried away by the prospects for algorithmic radiology, there is certainly a looming fear among radiologists that the computer overlords are warming up for some kind of takeover. Phelps Kelley, a fellow in radiology at UCSF, examined the state of the field and said, "The biggest concern is that we could be replaced by machines."[33] Radiology is one of the top-paid medical specialties, with annual compensation of about $400,000.[34] The performance of

Zebra Medical Vision's technology or the claim, advanced by Andrew Beam and Isaac Kohane, that computers are capable of reading
260 million scans in just twenty-four hours for only $1,000 makes
clear the economic reasons for replacing radiologists with machines.
Already, digitally outsourcing the radiological interpretation of images has become an increasingly popular cost-saving measure for
hospitals, with companies like vRad (Virtual Radiologic) employing
more than five hundred radiologists. Some 30 percent of American
hospitals use the service. Indeed, such outsourcing has grown expo-
nentially in recent years, and it is now the number one professional
service outsourced by hospitals. Hospitals are also training fewer
radiologists: the number of radiology residency slots offered in the
United States has dropped nearly 10 percent in the past five years
(although the number of radiologists in practice has been steadily
increasing, reaching just over 40,000 in 2016). Given the trends in
radiology, why not simply outsource to a machine?

At the moment, it still isn't possible. Gregory Moore, VP of
healthcare at Google and himself a radiologist, has observed that
"there literally have to be thousands of algorithms to even come close
to replicating what a radiologist can do on a given day. It's not going
to be all solved tomorrow."[35]

Earlier, I pointed out how difficult it has been to integrate all
the clinical data for each patient, even for the machine learning
companies and health systems that have both a patient's electronic
health records and the scan data. Making matters worse, any person who has received healthcare from many different providers over
a lifetime—which is to say, practically every American—defies a
capture of his or her comprehensive dataset. That provides major
challenges to computer-based image analysis.

Radiologists can provide a more holistic assessment than machines
can. Each scan is supposed to have a reason for being ordered, such
as "rule out lung cancer" for a chest X-ray. A narrow AI algorithm
could prove to be exceptionally accurate for ruling out or pointing toward the diagnosis of lung cancer. But, in contrast, the radiologist not

only scours the film for evidence of a lung nodule or lymph node enlargement but also looks for other abnormalities such as rib fractures, calcium deposits, heart enlargement, and fluid collections. Machines could ultimately be trained to do this, like researchers at Stanford working with their 400,000 chest X-ray collection, but deep learning of medical images so far has been fairly narrow and specific.

But even when that problem is solved, the answer to whether radiology should be given wholly over to machines is not as simple as the costs we pay in time or money. My experience reviewing my scans with my radiologist shows what the future of radiology could be. Even though I have chided those making wild-eyed predictions about the future of radiology, I do believe that eventually all medical scans will be read by machines. Indeed, as Nick Bryan asserted, "I predict that within 10 years no medical imaging study will be reviewed by a radiologist until it has been pre-analyzed by a machine."[36] To make sure that an image isn't only analyzed by a machine, however, radiologists need to change. As Jha and I wrote, "To avoid being displaced by computers, radiologists must allow themselves to be displaced by computers."[37]

If radiologists adapt and embrace a partnership with machines, they can have a bright future. Michael Recht and Nick Bryan, writing in the *Journal of the American College of Radiology* stated it well: "We believe that machine learning and AI will enhance both the value and the professional satisfaction of radiologists by allowing us to spend more time performing functions that add value and influence patient care and less time doing rote tasks that we neither enjoy nor perform as well as machines."[38] Although it might seem ironic, Yann LeCun, the computer scientist who is considered the founding father of convolutional neural networks, actually thinks that humans have a bright future in radiology as well. He thinks the simple cases will be automated, but that this won't reduce demand for radiologists. Instead, it will make their lives more interesting while enabling them to avoid errors that arise due to boredom, inattention, or fatigue.[39]

Not only could radiologists have more interesting professional lives, however, they could also play an invaluable role in the future of deep medicine by directly interacting with patients. They could be doing more for their patients, sometimes by doing less. As Marc Kohli, a UCSF radiologist, correctly pointed out, "We're largely hidden from patients. We're nearly completely invisible, with the exception of my name show[ing] up on a bill, which is a problem."[40] To start with, a very high proportion of scans are unnecessary or even downright inappropriate. Today, radiologists do not serve the role of gatekeepers; patients simply report for a scan, and it is performed by a technician. In the future, before an ordered scan is performed, the renaissance radiologist would review whether it is truly indicated and whether another type of scan might be better—such as an MRI or a CT for a question of a tear in the aorta. The radiologist would determine whether the scan ordered is necessary or of the right type and communicate the rationale to the patient.

This would have several benefits for the patient. The economic savings in avoiding wasteful imaging would be complemented by the reduced use of ionizing radiation, which over the course of a person's lifetime carries a cumulative risk of inducing cancer. This is where the partnership between radiologists and machines could bear even more fruit, as some encouraging studies have demonstrated the possibility of offsetting the reduced imaging power of lower doses of ionizing radiation by pairing those low-dose images with algorithms that greatly improve the quality of the image. Theoretically, with further refinements it will be possible to offer what are known as ultra-low dose CT scans, reducing the radiation by orders of magnitude and even lowering the cost of the CT machines themselves by eliminating the need for very high powered components. What an unexpected twist: machines disrupting machines instead of disrupting humans. Image enhancement algorithms are also being applied to MRIs with the goal of substantially reducing the amount of time it takes to scan a patient. Its inventors project a threefold improvement in efficiency, which would make it very attractive to clinics. But the biggest gains might just go to the patient, who would have to remain motionless while being subjected to loud banging noises in a claustrophobia-inducing tunnel for only ten minutes instead of sixty.[41] And all these benefits would be distinct from those derived from using AI to achieve improved interpretation of scans.

Besides gatekeeping, another important responsibility for radiologists would be discussing the results with the patient. That is now being done in some breast imaging centers, but it certainly has not been widely adopted across all types of imaging. The plusses for this are striking. Having some conversation and hearing more about the patient's symptoms or history helps the radiologist's assessment of the scan. It provides an independent point of view, distinct, as I've noted, from that of a surgeon, who is usually inclined to operate. The job of the master explainer is going to be especially vital with machine probabilistic outputs of scan findings. Here's a simple example of what an algorithm would produce: "Based on clinical and CT features, the probability that the nodule is lung cancer is 72 percent; the probability that it is benign is 28 percent." A typical response from a patient would be, "So, I have cancer." The radiologist could immediately mollify the patient's anxiety and explain that there is a more than one in four chance it won't turn out to be cancer.

The need for a human to integrate and explain medical results will become even more pronounced. Take the dreaded prediction of Alzheimer's disease. A group at McGill University developed and validated a deep learning algorithm in 273 patients who had a brain amyloid scan, APOE4 genotype, and follow-up clinical data. The algorithm had an accuracy of 84 percent for predicting Alzheimer's within two years.[42] Another poignant example is for longevity. A team led by Australian researchers used neural network analysis of 15,957 CT scans from individuals over the age of sixty to develop and validate a five-year survival plot, partitioning patients

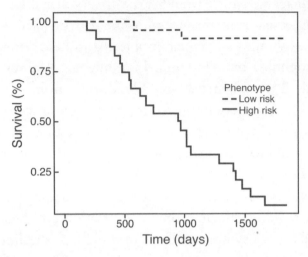

FIGURE 6.2: Predicting longevity from a deep neural network of CT scans. Source: Adapted from L. Oakden-Rayner et al., "Precision Radiology: Predicting Longevity Using Feature Engineering and Deep Learning Methods in a Radiomics Framework," *Sci Rep* (2017): 7(1), 1648.

by their risk of death, ranging from groups in which 7 percent were expected to die, to groups in which 87 percent were expected to die (Figure 6.2).[43] While today these algorithms are confined to research papers and haven't entered the realm of clinical care, it's just a matter of time until they are at least available for clinical care, even if they are not always applied. Among today's medical specialties, it will be the radiologist who, having a deep understanding of the nuances of such image-based diagnostic algorithms, is best positioned to communicate results to patients and provide guidance for how to respond to them. Nevertheless, although some have asserted that "radiologists of the future will be essential data scientists of medicine," I don't think that's necessarily the direction we're headed.[44] Instead, they likely will be connecting far more with patients, acting as real doctors.

For radiologists to be able to spend time with patients, serving as gatekeepers or master independent interpreters, their workload—the task of translating pixels into words—has to be decompressed. Already, AI algorithms can perform quantification and segmentation of images, which has eased a workflow burden for radiologists. Eventually, machines will take over the initial reading of an image and produce an initial draft of the scan report, with a radiologist's oversight required to sign off and make it official. Ideally, this will include data mining of the comprehensive medical information for each patient and its integration with the scan interpretation. That would save considerable time for the radiologist, who today often has to sift through the electronic medical record to try to connect the dots of the image with the patient's clinical background. We're likely some years away from that objective being comprehensively and routinely fulfilled.

EVEN BEFORE THE notion that machines might overtake doctors for reading scans, there were pigeons. A considerable body of data gathered over five decades has shown that pigeons can discriminate between complex visual stimuli, including the different emotional expressions of human faces as well as the paintings of Picasso and Monet. In 2015, Richard Levenson and colleagues tested whether pigeons could be trained to read radiology and pathology images.[45] The team placed twelve pigeons in operant conditioning chambers to learn and then to be tested on the detection of micro-calcifications and malignant masses that indicate breast cancer in mammograms and pathology slides, at four-, ten-, and twenty-times levels of magnification. Their flock-sourced findings were remarkably accurate. This led the researchers to conclude that one could use pigeons to replace clinicians "for relatively mundane tasks."

Machines are still more likely to supplement radiologists than birds are. After all, machines don't need food or cages. But, as the experiment with birds indicates, pathologists also face a future in which machines can learn to fill at least some of their roles.

Pathologists have varied roles and subspecializations. Some are into laboratory medicine, supervising clinical lab assays, others in forensics or autopsy. The pathologists in question here are those, such as surgical or cytopathology pathologists, who interpret slides of human tissue to make an unequivocal diagnosis of a disease. The problem: it's not so unequivocal.

There are innumerable studies showing striking heterogeneity of the interpretation—whether it's cancer, whether it's malignant, whether there's a transplant rejection, and on and on—of slides by pathologists. For example, in some forms of breast cancer, the agreement in diagnosis among pathologists can be as low as 48 percent.[46] Even with their training, length of experience, and subspecialty expertise, board-certified pathologists show a considerable error rate and are prone to overdiagnosis. There are multiple causes. Some of the diagnostic difficulty is related to the sample of tissue. Over the years, the trend has been to avoid surgery by using fine "skinny needle" aspirates, inserting a needle into an organ from outside the body. The technique has several advantages: the patient experience is better, and the cost of the procedure is lower because there's no need for an operating room, general anesthesia, or an incision. The problem is that there's often minimal tissue obtained. A suboptimal sample, which may not be representative of the tissue or organ under scrutiny, is not likely to be improved by machine processing, but many other aspects of diagnosis could be. One is a lack of standardization of techniques. There is also a throughput problem: a pathologist doesn't have an infinite amount of time to examine sets of slides with millions of cells.

In the analog days, pathologists had cabinets full of glass slides and had to peer through a microscope to look at each one. Now they look at a computer screen. Digital pathology has helped improve the workflow efficiency and accuracy of pathology slide diagnosis. In particular, the digital technique of whole slide imaging (WSI) enables a physician to view an entire tissue sample on a slide, eliminating the need to have a microscope camera attachment. Pathologists have been slower than expected to adopt WSI and other digital techniques, which in turn has slowed the encroachment of AI into pathology. Nevertheless, it's coming. WSI's most important feature going forward is that it provides the foundation for using neural network image processing in pathology.

The Stanford group used WSI to develop a machine learning algorithm to predict survival rates in patients with lung cancer, achieving accuracy better than current pathology practice using tumor grade and stage. Thousands of features were automatically identified from the images, of which 240 proved to be useful across non–small cell, squamous, and adenocarcinoma lung cancers.[47]

A number of other studies of deep learning for pathology interpretation have been encouraging, and many have been stimulated by an international competition known as the Camelyon Challenge. In 2016, Le Hou and the group at Stony Brook University used a convolutional neural network (CNN) to classify lung and brain cancer slide images at an accuracy level of 70 to 85 percent, with an agreement level similar to that of a group of community pathologists.[48] Google used high-resolution—gigapixel—images at forty-times magnification to detect metastasis with better than 92 percent accuracy, versus a 73 percent rate for pathologists, while reducing the false negative rate by 25 percent.[49] Google even gave the pathologists unlimited time to examine the slides. There were unexpected problems, however. The Google algorithm, regularly made false positive diagnoses. Similar issues were found in a large breast cancer detection deep learning study, with very few false negatives but far more false positives than a human would make.[50]

A key variable for pathologist accuracy turns out to be the amount of time allotted for reviewing slides. Another Camelyon Consortium report by Babak Bejnordi and colleagues assessed the performance of a series of algorithms against that of eleven pathologists in the detection of cancer spreading to lymph nodes.[51] When there were time limits imposed for the pathologists (less than one minute per slide, mimicking routine pathology workflow), the algorithms performed better. With unlimited time, the pathologist matched algorithmic accuracy.

As is the case with medical scans, the algorithms investigating pathology slides noticed things that can be missed by even expert human eyes, such as microscopic evidence of metastasis.[52] Likewise, deep learning can markedly improve the quality of microscope images, sidestepping the problems of out-of-focus or lower-quality slides.[53] And as is the case with medical imaging, algorithms can enhance, rather than replace, the human pathologist. The Computer Science and AI Laboratory (CSAIL) group at MIT developed a twenty-seven-layer deep network for diagnosis of cancer metastasis to lymph nodes with four hundred whole slide images.[54] The algorithm markedly reduced the pathologist error rate, but interestingly combining the pathologist and machine reading was clearly the best, with almost no errors. This complementarity of machines and humans, each making different correct and erroneous calls, along with neural networks optimizing quality of the slide images, is noteworthy. Synergy hasn't escaped notice of the multiple companies commercializing deep learning tools for path slide analytics (including 3Scan, Cernostics, Proscia, PathAI, Paige.AI, and ContextVision). For example, PathAI advertises an error rate with algorithms alone of 2.9 percent, and by pathologists alone of 3.5 percent, but the combination drops the error rate to 0.5 percent.

Pathologists do not only interpret slides. They can also examine samples at the molecular level, for example, by identifying the epigenetic methylation patterns on tissue DNA to improve cancer diagnosis. Like digital pathology and WSI, there's an overall lag of incorporating molecular diagnostics into routine pathologic assessment of cancer tissue. A study comparing machine analysis of methylation of brain cancer samples with pathologists' review of slides demonstrated the superiority of accuracy for algorithms when such methylation data was available.[55] In another study of pathology slides by researchers from New York University, the algorithmic accuracy for diagnosing subtypes of lung cancer was quite impressive (AUC = 0.97); half of the slides had been misclassified by pathologists. Further, their neural network was trained to recognize the pattern of ten common genomic mutations and predicted these from the slides with reasonable accuracy (0.73–0.86), especially for one of the early attempts to do so.[56] This finding is noteworthy because it exemplifies the ability of machine algorithms to see patterns not easily discernible by humans. As the incorporation of molecular diagnostics gets more common, including DNA sequence, RNA sequence, proteomics, and methylation, the advantage and complementarity of AI analytics, ingesting and processing large datasets, may become an especially welcome boon for pathologists.

Just as there is often marked disagreement for pathologists interpreting slides, there were differences in perception of the deep learning progress expressed in one of the major pathology journals. One group embraced machines:

> Computers will increasingly become integrated into the pathology workflow when they can improve accuracy in answering questions that are difficult for pathologists. Programs could conceivably count mitoses or quantitatively grade immunohistochemistry stains more accurately than humans, and they could identify regions of interest in a slide to reduce the time a pathologist spends screening, as is done in cytopathology. We predict that, over time, as computers gain more and more discriminatory abilities, they will reduce the amount of time it takes for a pathologist to render diagnosis, and, in the process, reduce the demand for pathologists as microscopists, potentially enabling pathologists to focus more cognitive resources on higher-level diagnostic and consultative takes (e.g., integrating molecular, morphologic, and clinical information to assist in treatment and clinical management decisions for individual patients).[57]

In contrast, the essay subtitled "Future Frenemies" called out the poor diagnostic accuracy of the deep learning algorithms to date, emphasizing the human edge: "We believe that a pathologic diagnosis is often a well-thought-out cognitive opinion, benefiting from our training and experience, and subject to our heuristics and biases."[58]

But it's more than human cognition. Much like radiologists, pathologists have had no real face to patients. It's their doctor who relays the report, and their doctor often has little appreciation for the nuances involved in the interpretation of a path specimen. Establishing direct patient contact to review the results could be transformative both for pathologists and for patients and their physicians.

The remarkable AI parallels for radiology and pathology led Saurabh Jha and me to write an essay in *JAMA* about the "information specialists."[59] Recognizing that many tasks for both specialties will be handled by AI and the fundamental likeness of these specialists, we proposed a unified discipline. This could be considered a natural fusion that could be achieved by a joint training program and accreditation that emphasizes AI, deep learning, data science, and Bayesian logic rather than pattern recognition. The board-certified information specialist would become an invaluable player on the healthcare team.

Tumor boards are a fine example. In modern practice, tumor boards are multispecialty groups that review each patient's diagnosis of cancer and treatment alternatives. Typically, this board includes a medical oncologist, a surgical oncologist, and a radiation oncologist, covering the gamut of drug, surgical, and radiation treatments available to a patient. But with the increasing prominence of AI in imaging and pathology, the information specialist will be an essential member of the team—the one who really understands the basis for deep learning diagnostic and prognostic algorithms. Of note, the first peer-reviewed research paper by IBM Watson Health was comparing its input with the molecular tumor board at the University of North Carolina Lineberger Comprehensive Cancer Center. Of more than a thousand cases retrospectively reviewed by the tumor board and Watson, more than 30 percent were augmented by AI information, particularly related to treatment options for specific mutations.[60]

LIKE RADIOLOGY AND pathology, dermatology involves a great deal of pattern recognition. Skin conditions are among the most frequent reasons for seeing a doctor—they account for 15 percent of all doctor visits! Unlike radiology and pathology, however, about two-thirds of skin conditions are diagnosed by non-dermatologists, who frequently get the diagnosis wrong: some articles cite error rates as high as 50 percent. And, of course, dermatologists don't just look at and diagnose skin rashes and lesions, they often treat or excise them. But the pattern recognition of skin problems is a big part of medicine and a chance for artificial intelligence to play a significant role. With relatively few practicing dermatologists in the United States, it's a perfect case for machines to kick in.

The digital processing of smartphone selfie skin lesions got off to a rocky start with a proliferation of mobile apps with variable results. Back in 2013, an evaluation of smartphone apps for melanoma diagnostic accuracy showed that cancerous growths were wrongly classified as benign 30 percent of the time.[61] Accuracy ranged wildly from 6.8 to 98.1 percent. Another study assessing three apps showed poor sensitivity (21 to 72 percent) and highly variable specificity (27 to 100 percent) compared with dermatologists.[62]

The patterns that a dermatologist might need to recognize are a broad group, including skin rashes and lesions, but correctly identifying skin cancer is regarded as the principal goal for AI in dermatology. This is the case especially for the early detection of melanoma, before it spreads to lymph nodes and throughout the body, as that translates to a far better five-year survival rate (99 percent if detected early versus 14 percent if detected in late stages).[63]

Overall, skin cancer is the most frequent human malignancy, with its highest incidence in Australia and New Zealand (about 50 per 100,000 population) and 30 per 100,000 in the United States. This translates to more than 5.4 million Americans with new cases of skin cancer each year, at a cost of over $8 billion. One in five Americans will develop skin cancer in their lifetime. But, fortunately, nonmelanomas are twenty times more common than melanoma. The critical issue is

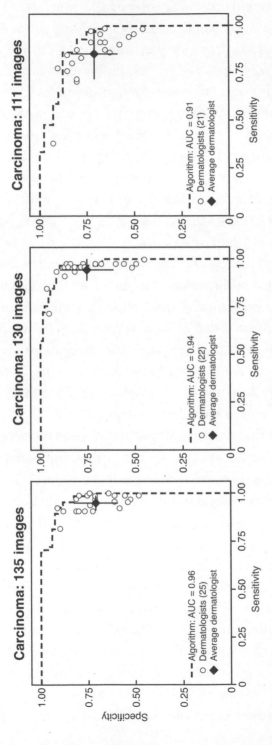

FIGURE 6.3: Skin cancers classified by a deep learning algorithm and dermatologists. For each classification, the algorithm performed at least as well as the group of more than twenty dermatologists. Source: Adapted from A. Esteva et al., "Dermatologist-Level Classification of Skin Cancer with Deep Neural Networks," *Nature* (2017): 542(7639), 115–118.

to distinguish between keratinocyte carcinomas, the most common skin cancer, with a very high cure rate, versus malignant melanoma. Wrongly identifying a growth as melanoma can lead to unnecessary biopsies (especially among non-dermatologists) for innocent, benign lesions. Missing one is worse: melanoma kills approximately 10,000 Americans each year.

The classic way dermatologists diagnose melanoma is using a heuristic, an acronym known as ABCDE, which stands for asymmetry, irregular border, more than one or uneven distribution of color, a large (greater than 6 mm) diameter, and evidence that the mole is evolving. Dermatologists not only rely on their experience and eye but use a dermoscope to magnify and light up the lesion of interest. That's very different from photographs of a skin lesion taken at variable distances and angles with different lighting conditions. The looming question in the era of AI was whether this could be simulated or even exceeded with deep learning.

In 2017 one of the most impressive deep learning papers yet published appeared in *Nature*, with its "Lesions Learnt" cover, on the diagnosis of skin cancer.[64] The algorithm had two goals: accurately classify lesions as benign or malignant, and if malignant, whether it is melanoma. Andre Esteva and colleagues at Stanford University used a Google CNN algorithm (GoogleNet Inception v3) that was pretrained with ImageNet's 1.28 million nonmedical images of more than one thousand object classes. The neural network was trained via 129,450 images of skin lesions, representing 2,032 skin diseases (Figure 6.3). But many of these images were photographs, not biopsies, so definitive validation was performed with biopsy-based diagnosis on 1,942 lesions, resulting in a yes/no classification both of whether the image showed cancer and whether the cancer was malignant (including both photographs and dermoscope images for the latter task). The results were tested against more than twenty board-certified Stanford dermatologists. Each dermatologist, who had not seen any of the lesions previously, was asked whether he or she would biopsy the lesion or reassure the patient. The algorithm outperformed each of them for classifying cancer in 135 dermo-scopic images and was better than the average dermatologist for 130 photographic images of melanoma and 111 dermoscopic images of melanoma (Figure 6.3).

The Stanford paper's use of a CNN algorithm for assessing skin cancer was replicated by IBM Watson with higher accuracy com-pared with eight expert dermatologists for melanoma.[65] A subsequent refinement of the algorithm used by the Stanford study (Google's Inception v4 CNN) was put to the test against a much larger group of fifty-eight dermatologists for the specific diagnosis of melanoma, and again most humans were outperformed.[66]

There are major implications and questions about this research. The accompanying *Nature* editorialists asked whether medical staff would "become mere technicians responding to a machine's diagnos-tic decisions."[67] Those are precisely the broad implications that this book is meant to address, but the editorial also recognized, critically, that an algorithmic test drive is not the same as using a technology in the real world. To date, very few patients of non-European ancestry have been included in algorithmic training.[68] The CNN has to be clinically validated. That's why it was surprising to see one of the skin lesion studies release its algorithm to the public for mobile device use without such validation.[69] Dermatologists assessed in these studies were not looking at patients and didn't have to worry about making an incorrect diagnosis. In the real world of dermatology, it isn't just a look at a lesion that is telling. The history of the lesion, the indi-vidual's risk factors, a more extensive assessment of the whole skin of the patient are all in the domain of information the dermatologists acquire during a visit. Furthermore, it's not just a binary call of can-cer or not cancer but the decision to simply monitor the lesion over time before deciding to do a biopsy. Accordingly, we can consider the algorithm as a contrived, narrow way to make a diagnosis and biopsy plan compared to the real, clinical world. Nonetheless, it's clear that deep learning can help promote accuracy in detection of skin cancer. What the Stanford study did show unequivocally was that it's ready for further testing. Other algorithms are in the works, too. Companies like VisualDx, which has teamed up with Apple's machine learning group, are helping to make diagnoses of skin rashes and lesions well beyond cancer from a database of over 30,000 images.[70] These collective efforts could ultimately have implications anywhere in the world, anytime, where there is a person with a smartphone, broadband Internet access, and a skin lesion of interest.

As I noted, there are not very many dermatologists in the United States: fewer than 12,000 dermatologists to look after more than 325 million Americans. So the story here isn't so much replacing dermatologists with machines as empowering the family physicians and general practitioners who are called on to do most of the dermatological grunt work. A fully validated, accurate algorithm would have a striking impact on the diagnosis and treatment of skin conditions. For dermatologists, it would reduce the diagnostic component of their work and shift it more to the excision and treatment of skin lesions. It would make primary-care doctors, who are the main screening force for skin problems, more accurate. For patients, who might otherwise be subject to unnecessary biopsies or lesion removals, some procedures could be preempted.

From this tour of medical scans, path slides, and skin lesions, we've seen the potential role of AI to change medicine by ameliorating the accuracy of diagnosis and making the process more efficient. This is the "sweet spot" of AI, reading patterns. But certainly no data have yet emerged about replacing doctors, even in these most "vulnerable" specialties. Now let's turn to clinicians who are not looking at such classical patterns in their daily practice.

chapter seven

CLINICIANS WITHOUT PATTERNS

> Medical diagnostic AI can dig through years of data about cancer or diabetes patients and find correlations between various characteristics, habits, or symptoms in order to aid in preventing or diagnosing the disease. Does it matter that none of it "matters" to the machine as long as it's a useful tool?
>
> —GARRY KASPAROV

UNLIKE INFORMATION SPECIALISTS, MOST PHYSICIANS, NURSES, and clinicians do not have what I'd call pattern-centric practices. Most primary care and specialties involve pattern-heavy elements such as scans or slides, to be sure, but their predominant function is making an assessment and formulating a plan. This involves integration; cognitive processing of the patient's history; physical, lab, and other objective data (such as the scans and slides interpreted by information specialists); the corpus of the medical literature; and communication with the patient and family. The practice of these clinicians defies a simple pattern. While deep learning thrives on inputs and outputs, most of medical practice defies straightforward algorithmic processing. For these "clinicians without patterns," AI presents adjunctive opportunities, offsetting certain functions that are more efficiently handled by machines. The range is exceptionally broad—from eliminating the use of a keyboard to processing multi-modal data.

One area in which AI was touted early and heavily was digesting the enormous research output in biomedicine. Each year more than 2 million peer-reviewed papers are published—that's one article every thirty seconds—and there isn't one human being who could keep up with all this new information, much less busy doctors. It was a bit comical in 2017 to see IBM Watson's ads claiming that with the system a doctor could read 5,000 studies a day and still see patients. Neither Watson nor any other AI algorithms can support that, at least not yet. What Watson is actually dealing with are only abstracts, the brief encapsulations that appear at the beginning of most published papers. Even then, these data are unstructured, so there's no way that simple ingestion of all the text automatically translates into an augmented knowledge base.

This might be surprising, given that Watson's ability to outdo humans in *Jeopardy!* suggests it would have the ability to outsmart doctors, too, and make quick work of the medical literature. It turns out all Watson did to beat humans in the game show was to essentially ingest *Wikipedia,* from which more than 95 percent of the show's questions were sourced. Gleaning information from biomedical literature is not like making sense of *Wikipedia* entries. A computer reading a scientific paper requires human oversight to pick out key words and findings. Indeed, Andrew Su, a member of our group at Scripps Research, has a big project called Mark2Cure using web-based crowdsourcing, with participants drawn from outside the scientific community, to do this work. Volunteers (we call them citizen scientists) mine and annotate the biomedical literature, as represented by the more than 20 million articles in PubMed, a research database run by the National Institutes of Health. No software today has the natural-language-processing capabilities to achieve this vital function. But it is certainly coming along. At some point in the years ahead, Watson might live up to its hype, enabling all doctors to keep up with the medical literature relevant to their practice, provided it was optimally filtered and user friendly.

In the old days before electronic medical records, a patient's complexity could be quickly assessed by a symptom known as the thick chart sign. Despite the advent of electronic medical records, the thick chart is still a frequent phenomenon for many new patients or second opinion visits, as copies of records come in via fax or e-mail, often containing tens or hundreds of pages to review.

Digital record keeping was supposed to make the lives of clinicians much easier. But the widely used versions of electronic records defy simple organization or searchability and counter our ability to grasp key nuggets of data about the person we're about to see. Just the fact that it takes more than twenty hours to be trained to use electronic healthcare records (EHR) indicates that the complexity of working with them often exceeds that of the patient being assessed. Perhaps even worse than poor searchability is incompleteness. We know there's much more data and information about a given person than is found in the EHR. There are medical encounters from other health systems and providers. There are antecedent illnesses and problems when the individual was younger or living in another place. There are data from sensors, like blood pressure, heart rhythm, or glucose, that are not entered in the chart. There are genomic data that millions of people have obtained that are not integrated into the record. And there is the social media content from networks like Facebook that goes ignored, too. Even if clinicians could work well with a patient's EHR, it still provides a very narrow, incomplete view.

Electronic health records also defy the power of AI tools. As with the relevant medical literature, ideally AI could mine and integrate all the data about a patient, if they were structured comprehensively, neatly, and compactly. We haven't seen such a product yet. If we ever do, it will represent not only a means to promote the efficiency of the doctor's workflow but also a tool for a more meaningful and thorough

FIGURE 7.1: The disconnected doctor and patient. Source: Adapted from "The Pharos," *The Pharos of Alpha Omega Alpha Honor Medical Society,* Summer Edition, 78 (2015).

assessment of each patient. And it will ultimately prove to be immensely valuable to each individual in his or her life journey through health and illness.

If there's one thing that both doctors and patients hate about an office visit, it's the use of keyboards (Figure 7.1). Pecking away at a keyboard distracts the doctor and disengages the patient. Face-to-face contact, the opportunity to take in body language, and the essence of interpersonal communication are all lost. It goes both ways: the patient has no sense of whether the doctor has any empathy; doctors, frustrated by the burden of electronic charting, know full well that their listening and engagement abilities are compromised. This modern e-ritual has contributed to the peak incidence of burnout and depression seen among physicians.

When electronic health records first arrived, a new industry of human scribes was birthed to preserve the human interaction between patient and doctor by outsourcing the keyboard function to a third party. ScribeAmerica is now the largest of more than twenty companies providing transcription services to health systems and clinics throughout the United States. By 2016, there were more than

20,000 scribes employed and a projection that there would be a demand for more than 100,000 by 2020—that's one scribe for every seven physicians.[1] Having a scribe at the keyboard has been associated with multiple reports of improved satisfaction for both patients and doctors. But it's hardly a full remedy. Adding a substantial number of full-time employees exacerbates the exorbitant costs of electronic health information systems such as Epic or Cerner. Beyond that, the presence of an extra, unfamiliar person in the room can interfere with a patient having an intimate conversation with a physician.

Adding computers to the doctor's office has thus far represented the primary attempt to digitize medical practice, and many consider them an abject failure. Ironically, perhaps, there may be a machine solution to the problem. In a world of voice assistants like Alexa, it makes sense to ask whether typing is even necessary. Speech is so much faster. And entry of data to the unwieldy EHR takes inordinate time. For example, to enter the smoking history of a patient— say, he smoked three packs of cigarettes a day for twenty years and stopped smoking five years ago—could take several minutes. But it would take just a couple of seconds to say it.

You would think this could be a layup for AI. AI speech processing already exceeds the performance of human transcription professionals. Why not have the audio portion of the visit captured and fully transcribed, and then have this unstructured conversation synthesized into an office note? The self-documented note could be edited by the patient and then go through the process of both doctor review and machine learning (specific to the doctor's note preferences and style). After fifty or more notes processed in this way, there would be progressively less need for a careful review of the note before it was deposited into the electronic record. This would make for a seamless, efficient way of using natural-language processing to replace human scribes, reduce costs, and preserve face-to-face patient-doctor communication.

Further, by enlisting patients in editing their notes, some of the well-recognized errors that plague medical visits and electronic health records could be addressed. Having the entire visit taped and potentially transcribed would create an archive that patients would later have the opportunity to review—especially useful as so much of what is discussed in an appointment will not necessarily be fully understood or retained by patients. We know that 80 percent of office notes are cut and pasted, with mistakes propagated from one visit to another, and from one doctor to another.[2] Often medications are listed that are not active (or were never even prescribed) or medical conditions are included but are not accurate. The patient's input has not been previously solicited, but it could be very helpful in cleaning the data. Some physicians remain concerned that this would lead to new inaccuracies, but the trade-off compared with our current state may well be beneficial. We'll see: a digital scribe pilot combining natural-language processing (to transcribe the speech from the visit) and machine learning (to synthesize the note) has been initiated at Stanford University in conjunction with Google, and work to develop algorithms to produce such visit notes are being actively pursued by companies like Microsoft, Amazon, Google, Nuance, and many start-ups, such as Sopris Health, Orbita, CareVoice, Saykara, Augmedix, Sensely, Suki, and Notable.[3]

The NLP-visit record might still prove to be suboptimal. Beyond the technical challenge of transcribing unstructured language into a succinct but complete note, there are the missing pieces. All the non-verbal communication would be lost, for example. The knowledge that everything was going to be taped, archived, and parts would be incorporated into the note might well inhibit an informal, free conversation. AI is actively being pursued for this application despite uncertainty that it will ultimately prove widely acceptable.

Other components of a doctor's visit are well suited for machine learning. Artificial intelligence already forms the backbone of tools known as clinical decision support systems (CDSS). These algorithms, which have been used and updated over the past couple of decades, were supposed to provide a range of functions to make the doctor's job easier and improve quality of care: reviewing the patient's data; suggesting diagnoses, lab tests, or scans; recommending vaccinations; flagging drug allergies and drug interactions; and avoiding potential medication errors. They haven't quite lived up to the billing so far: A systematic review of twenty-eight randomized trials of CDSS failed to show a survival benefit but did show a small improvement in preventing morbidity.[4] A major concern about CDSS to date has been disruption of workflow, with too many prompts and alerts. Beyond that, current decision support systems are primitive compared to recent progress made in AI. One thing that will improve them is the ability to ingest all the medical literature. Even though it isn't possible yet, eventually that will be accomplished, bringing a vast knowledge base to the point of care of individual patients and facilitating medical diagnoses and optimal treatment recommendations. That ought to be far better than the standard today, which is for doctors to look things up on Google or, less frequently (due to cost of access), checking for recommendations at UpToDate, software that compiles medical evidence and is embedded in the CDSS of some health systems.

Up-to-the-minute biomedical research would be useful, but it isn't the goal. Ralph Horwitz and colleagues wrote a thoughtful perspective, "From Evidence Based Medicine to Medicine Based Evidence," that quoted Austin Bradford Hill, an eminent English epidemiologist, on what doctors weren't getting from research. "It does not tell the doctor what he wants to know," Hill said. "It may be so constituted as to show without any doubt that treatment A is on the average better than treatment B. On the other hand, that result does not answer the practicing doctor's question: what is the most likely outcome when this particular drug is given to a particular patient?"[5]

To make the best decision for a particular patient, a physician or an AI system would incorporate all of the individual's data—biological, physiological, social, behavioral, environmental—instead of relying on the overall effects at a large-cohort level. For example, extensive randomized trial data for the use of statins shows that for every one hundred people treated, two to three will have a reduction in heart attack. The rest will take the drug without any clinical benefit besides a better cholesterol lab test result. For decades, we've known the clinical factors that pose a risk for heart disease, like smoking and diabetes, and now we can factor in genetic data with a risk score from an inexpensive gene array (data that can be obtained for $50 to $100 via 23andMe, AncestryDNA, and other companies). That score, independent of as well as in addition to traditional clinical risk factors, predicts the likelihood of heart disease and whether use of a statin will benefit that individual. Similar genetic risk scores are now validated for a variety of conditions including breast cancer, prostate cancer, atrial fibrillation, diabetes, and Alzheimer's disease.

Drilling down on data with smart AI tools would also include processing an individual's labs. Today, lab scores are also considered against population-based scales, relying on a dumbed-down method of ascertaining whether a given metric is in a "normal" range. This approach reflects the medical community's fixation on the average patient, who does not exist. For example, there's no consideration of ancestry and ethnic specificity for lab tests, when we know that key results—such as hemoglobin A1C, which is used to monitor diabetes, or serum creatinine, which is used to monitor kidney function—are very different for people of African ancestry than for those of European ancestry.[6] Moreover, plenty of information is hidden inside the so-called normal range. Take a male patient whose hemoglobin has steadily declined over the past five years, from 15.9 to 13.2 g/dl. Both the starting and end points are within the normal range, so such a change would never get flagged by lab reports, and most busy doctors wouldn't connect the dots of looking back over an extended period. But such a decrement could be an early sign of a disease process in the individual, whether it be hidden bleeding or cancer. We have trapped ourselves in a binary world of data interpretation—normal or abnormal—and are ignoring rich, granular, and continuous data that we could be taking advantage of. That's where deep learning about an individual's comprehensive, seamlessly updated information could play an important role in telling doctors what they want to know. Instead of CDSS, I'd call it AIMS, for augmented individualized medical support.

So far, we've reviewed where AI could have generalized impact for all doctors and clinicians. Now let's look at some specialties for which there are AI initiatives or results to gauge progress. None of these have yet been implemented in routine medical practice, but they are good indicators for where the field is headed.

EYE DOCTORS

Although radiology and pathology are specialties where AI is making early and rapid progress, the exceptional recent progress in diagnosing diseases of the eye with AI makes me think this could turn out to be the pacesetter over time.

The number one global cause of vision loss is diabetic retinopathy, affecting more than 100 million people worldwide. Its prevalence in the United States is estimated to be nearly 30 percent of people with diabetes.[7] Diabetic retinopathy is an enormous public health issue for which routine screening is recommended, but more often than not it doesn't get done despite effective treatments that can delay progression and prevent blindness.

If all the recommended screening of people with diabetes was performed, there would be well over 30 million retinal images per year that would need to be evaluated.[8] Clearly, this sounds like a job for deep learning. A group led by researchers at Google has developed an algorithm to automatically detect diabetic retinopathy and diabetic macular edema.[9] Very little technical information was provided about the convolutional neural network used, besides a reference to the paper by Christian Szegedy and colleagues at Google and the Inception-v2 architecture.[10] What we do know is that they used 128,175 retinal images for training and two sets of validation images (9,963 and 1,748), cumulatively involving over 75,000 patients. The retinal images were also graded by more than sixty board-certified ophthalmologists, some of whom read thousands of images (median range from 1745 to 8906). The software developed had an impressive sensitivity of 87 to 90 percent and 98 percent specificity.[11] Google wasn't the only team to develop a deep learning algorithm for diabetic retinopathy. Using more than 35,000 retinal images, IBM reported an 86 percent accuracy.[12] Even a sixteen-year-old, Kavya Kopparapu, developed such an algorithm by adopting Microsoft's ResNet-50 and using training data from 34,000 images available from the National Eye Institute. She, her brother, and her team formed the company Eyeagnosis and developed a 3-D-printed lens attachment for a smartphone so that the algorithm could potentially be used to diagnose diabetic retinopathy anywhere.[13]

There are a couple of factors to keep in mind regarding these encouraging findings: diabetics with retinopathy are more likely to have poorer pupil dilation and more severe cataracts, both of which could obscure adequate images for the algorithm. Furthermore, the results may be influenced by who is using the retinal camera—this could include not just ophthalmologists but optometrists and other clinicians. These questions were addressed in the first prospective clinical trial of artificial intelligence in medicine. A spinoff company of the University of Iowa ophthalmology group named IDx developed a deep learning algorithm that used a Topcon retinal camera to detect diabetic retinopathy. At ten sites in the United States, a group of nine hundred patients with diabetes prospectively underwent eye exams with the IDx machine plus algorithm in the offices of primary-care doctors. Images were instantly transferred to the cloud for analysis, and the results were available within a few minutes. The accuracy for diabetic retinopathy was high, with a sensitivity of 87 percent and specificity of 90 percent.[14] Note that this prospective study, a first in clinical trials of its kind, did not have nearly the same level of accuracy as retrospective reports (compared with AUC = 0.99 in two such datasets using different algorithms). IDx was approved by the FDA in 2018. While uptake may be limited because it requires the IDx system, which costs more than $20,000, nonetheless, the technology represents a step toward accurate machine diagnosis of this condition without the need for ophthalmologists, as my colleague Pearse Keane and I wrote in the accompanying editorial for this important report.[15]

The other prominent cause of blindness is age-related macular degeneration (AMD), and, as with diabetic retinopathy, timely treatment can often prevent or at least delay the disease. In 2018, I visited the Moorfields Eye Hospital in London, one of the most highly regarded eye centers in the world. Pearse Keane, a pioneering ophthalmologist there, performed my eye exam with optical coherence tomography (OCT) (Figure 7.2). As opposed to head-on images of the retinal fundus, which were used in the diabetic retinopathy studies, the OCT images are cross sections of the retinal tissue. This sounds pretty fancy, but the images were simply obtained by resting my head on a machine and having my retina scanned by an array of light, one eye at a time. The image was ready to review in less than a minute. Keane developed a collaboration with DeepMind AI to build a deep learning algorithm that could help with more than 1 million high resolution 3-D OCT exams per year that are done at Moorfields. The algorithm can accurately diagnose most retinal diseases, including age-related macular degeneration, well before symptoms may be present. In a collaborative study between Moorfields and DeepMind, involving more than 14,000 OCT images, automated OCT interpretation was at least as accurate as expert retinal specialists for analysis and triage for urgent referral of more than fifty types of eye diseases, including glaucoma, diabetic retinopathy, and age-related macular degeneration.[16] There was not a single case in which the patient had a serious eye condition, but the algorithm recommended just observation. The algorithm's AUC for false alarm was 0.992. In contrast, the clinicians only agreed on 65 percent of the referral decisions. Keane told me that OCT should be part of every eye exam. That is certainly not the case in the United States today, but the prospective validation of the algorithm, which is currently underway in a clinical trial led by Keane, may lead to that standard practice in the future. And the accuracy for urgent referral

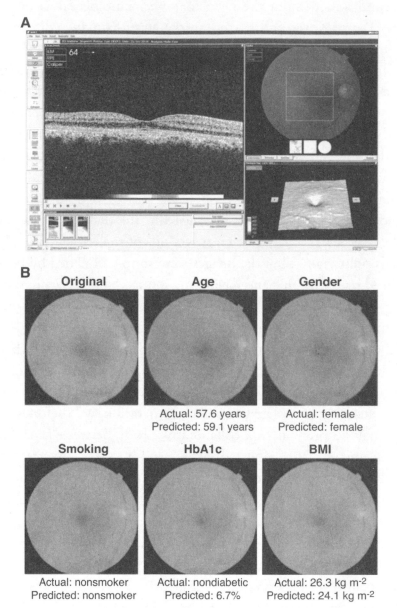

FIGURE 7.2: Images of the retina. (A) My optical coherence tomography (OCT) exam at Moorfields Eye Hospital. (B) retinal images that predict key metrics. Source: Adapted from R. Poplin et al., "Prediction of Cardiovascular Risk Factors from Retinal Fundus Photographs via Deep Learning," *Nature Biomedical Engineering* (2018): 2, 158–164.

of eye conditions will likely be substantially improved with the deep neural network his team developed.

Similarly, an OCT interpretation algorithm developed by Kang Zhang and his colleagues at UCSD, with almost 110,000 images, compared favorably to performance by ophthalmologists for accurately diagnosing AMD.[17] Zhang and others are working on a smartphone attachment that could achieve the image capture similar to specialized OCT machines.

Neural networks of retinal images can provide far more than just AMD insights. As we learned from a Google study of more than 300,000 patients, retinal images can predict a patient's age, gender, blood pressure, smoking status, diabetes control (via hemoglobin A1c), and risk of major cardiovascular events, all without knowledge of clinical factors (Figure 7.2).[18] The predictive accuracy was quite good for age and gender, and moderate for smoking status, blood pressure, and adverse outcomes. Such a study suggests the potential of a far greater role for eyes as a window into the body for monitoring patients. If such approaches are prospectively validated, we might see wide use of periodic retina self-exams with a smartphone in the future. This might capture control of blood pressure and diabetes and risk of outcomes, along with early diagnosis and tracking of AMD, diabetic retinopathy, glaucoma and cataracts, and even early signs of Alzheimer's disease.[19] It might even be extended to accurate eye refraction for updating eyeglass prescriptions. People don't like to put in drops to dilate their pupils, which could be a limitation for the popularity of smartphone selfie eye exams. But there are possible workarounds such as using infrared light. The combination of low cost, ease of (noninvasive) image capture, and wealth of data might someday be transformative.

There are also opportunities to help children and their vision with AI. One addresses a very challenging diagnosis: retinopathy of prematurity. This condition affects two of three preemie babies who weigh less than 1,250 grams at birth. It is often missed; neonatologists are not pediatric ophthalmologists, the experts in making this important diagnosis, and the exam in the neonatal intensive care unit is subjective and suboptimal. It is crucial that we do better, as retinopathy of prematurity is a leading cause of childhood blindness but is also a treatable condition. Deep learning has been shown to be exceptionally accurate, as good or better than retinopathy of prematurity experts, in a large retrospective study of 6,000 images.[20]

The other condition in kids that can be helped by AI is congenital cataracts, which would ordinarily be diagnosed and managed in specialized centers.[21] Congenital cataracts are much more complex than typical senile cataracts with respect to classifying images of the lens, making an accurate diagnosis, and making the best decision about surgery. As with diagnosing diabetic retinopathy, the illumination intensity, angle, and image resolution vary across imaging machines and doctors, creating diagnostic challenges. A study in China used ophthalmologist-labeled images from 886 patients to train a deep learning network called the CC-Cruiser agent. Using 410 eye images from children with congenital cataracts and 476 normal images, a seven-layer CNN was used that was derived from ImageNet, analyzing 4,096 features. The neural network prospectively diagnosed all but one of 57 patients with this rare condition accurately and provided treatment decisions in a multihospital clinical trial in China, with similar findings in an additional collaborative cloud-platform, website-based study. This backed up the claim that the algorithm was capable of expert ophthalmologist performance. More broadly, the implications of such a pioneering effort for rare diseases are notable, making the potential for deep learning algorithms to be useful outside of specialized referral centers. Besides promoting access, the global pooling of such data could be quite useful for ameliorating future performance of AI algorithms for congenital cataracts.

HEART DOCTORS

There are many subtypes of cardiologists: general, interventional (also known as plumbers, they unclog arteries), electrophysiologists (also known as electricians, they treat heart-rhythm disturbances), imaging specialists (more like radiologists), and heart failure experts. So, there is not one principal activity or function that is the same across the board, but there are two fundamental technologies— electrocardiography (ECG) and echocardiography (echo)—that are heavily relied upon by all.

ECGs have been read by machines for decades. An ECG has twelve leads, of which six track different vectors of the heart's electrical activity (classically obtained from electrodes placed on each of the four extremities), and six leads applied directly on the chest at different, standardized locations. Automated systems were first applied to reading ECGs in the 1970s and became routine by the 1980s, an adoption milestone that may be considered the first major entry of AI into the practice of medicine. But it certainly wasn't AI by today's standards. In fact, back then, we didn't hear about artificial intelligence. The term was simply "computer aided."

When I was on my cardiology rotation in 1981, in the midst of internal medicine training at UCSF, I'd go down to the office of Melvin Scheinman, the senior electrophysiologist, each day to read a stack of forty to fifty ECGs sitting in a wire letter basket. Each had a computer diagnosis printed on it. I was supposed to read the ECG without looking at that, but it was hard to not take a glance at what the machine had to say. It was especially fun to find the mistakes, which were not infrequent. That still is the case today because AI algorithms used for ECG interpretation are not very smart. They don't learn. Instead, they are heuristic, based on static rules for distinguishing patterns. ECG algorithms were assessed in 1991 in a large international study. The overall accuracy was 69 percent.[22] The same algorithms are still routinely used today throughout US hospitals and clinics.

Indeed, it is surprising how little has been done with modern AI tools to improve automated ECG interpretation accuracy. A neural network used to diagnose heart attacks, published in 1997, had one input layer, one hidden layer of fifteen neurons, and one output layer.[23] Adding more neurons in the single hidden layer helped improve accuracy.[24] But one hidden layer is not deep. Like all of its twelve-lead ECG machine reading precedents, it was rules based. The net result today is that with more than 300 million ECGs done per year, after forty years there must have been tens of billions of ECGs read via rules-based algorithms. The same algorithms are used to read the multiple twelve-lead ECGs that are obtained during a treadmill stress test. It seems, at least until now, that the companies involved in acquiring ECGs have not been motivated to improve machine accuracy, leaving a noteworthy human edge for one of the first and durable entries of AI into medicine. It's one of the reasons that I still love to read ECGs with students and trainees, to make sure they never trust the machine's diagnosis.

With the recent development of a DNN algorithm that fairly accurately diagnoses heart attack (sensitivity 93 percent, specificity 90 percent) by the twelve-lead ECG, we may be starting to get past the stagnation with rules-based interpretation.[25] In contrast to the twelve-lead ECG, there has been progress in diagnosing heart rhythm via a single lead using modern deep learning approaches. That was, in part, made possible by better technology for continuously recording a person's heart rhythm. The standard tool is the Holter monitor, invented by Norman Holter in 1949, despite needing to wear multiple electrodes with wires attached. The iRhythm Zio patch, which I described early in the book, is a Band-Aid that is mailed to patients for diagnosing heart rhythm. It is placed on the chest, records a single lead, and can typically capture all heartbeats for ten to fourteen days without impeding exercise or showering. The advance in technology created a dataset at iRhythm that was five hundred times larger than anything previously studied for heart-rhythm determination. A thirty-four-layer CNN was used by the Stanford University group, led by Andrew Ng, to analyze 64,121 thirty-second ECG strips from 29,163 patients, with ground truths established by technicians certified in ECG reading.[26] Testing then proceeded with 336 ECG records from 328 patients, and the algorithm was compared with readings by six board-certified cardiologists (and three others for ground truth labels). In all, twelve different abnormal rhythms were diagnosed, including atrial fibrillation and heart block, along with normal sinus (our normal pacemaker) rhythm. In this retrospective study, the algorithm outperformed the six cardiologists for most of the arrhythmia categories. But there were mistakes by both machines and heart doctors with aggregate positive predictive values in the 70 to 80 percent range.

Diagnosing atrial fibrillation is especially important. It's common, with an approximate 30 percent lifetime risk in the general population; it often happens without symptoms; and it carries an important risk of stroke. In late 2017, the FDA approved a system developed by AliveCor that combines a watchband-mounted sensor with a deep learning algorithm for diagnosing atrial fibrillation. The ECG single-lead sensor, which monitors heart rate continuously when it is worn, connects with the Apple Watch. A user can generate a thirty-second ECG at any time by placing a thumb on the watchband. The ECG is analyzed by an algorithm just like the smartphone sensor attachment that has been in use for more than five years. AliveCor also took advantage of the accelerometers that track a person's movement to identify heart-rhythm disturbances by identifying heart rates that were disproportionate to activity levels. An unsupervised learning neural network runs every five seconds and predicts the relationship between an individual's heart rate and physical activity. A nonlinear pattern is evidence of discordance, prompting the device to alert the user to take an ECG during the precise temporal window when a possible arrhythmia is occurring. Unlike all the other technologies I've reviewed so far, this tool is intended for consumers, not doctors—the whole point is to record the patient's activity in the real world, not a doctor's office, after all. But strips that are recorded, archived, and easily sent to a heart (or any) doctor may be quite helpful in making a diagnosis.

Temporal disconnects between the acquisition of an ECG and the heart rate–physical activity discordance might explain struggles with similar technologies. Another company that has worked on the smartwatch for heart-rhythm diagnostics is Cardiogram. More than 6,000 people wore the Apple smartwatch with their Deep Heart machine learning algorithm app for an average of almost nine weeks, but the accuracy for atrial fibrillation was poor: only 67 percent sensitivity and specificity.[27]

Echocardiography, cardiology's other vital technology, is used to evaluate heart structure and function. The motion of the heart, along with precise definition of key structures in echo loops like the endocardium—the inner layer of the heart—makes it difficult for a complete, automated edge-detection analysis, but there are nonetheless ongoing efforts to deploy AI tools to process echocardiographic exams, such as at UC Berkeley and Ultromics, a start-up company in Oxford, England (a spin-off of Oxford University).[28] The Berkeley group published the first deep neural network applied to echocardiography, comparing the machine image interpretation with the work of board-certified cardiologists at UCSF. While the retrospective study was relatively small, with only a few hundred patient images, the accuracy was quite good, with the algorithm capturing more than 90 percent of what the cardiologists saw.[29] Ultromics is focused on stress-echo image interpretation, which refers to the comparison of the echocardiogram at baseline, before any exercise, and the one obtained at peak exercise. The company claims on its website an accuracy of over 90 percent for the diagnosis of coronary artery disease, but data have not yet been published.[30] One of the new smartphone ultrasound companies, Butterfly Network, has used AI to detect the ultrasound probe's position and image output, suggesting adjustment of the probe's position through a deep learning algorithm. Given echo's prominent role in cardiac diagnosis and management, far more AI work needs to be done for automating its interpretation. Rapid, accurate machine analysis would be useful for most doctors, since they are not trained in echocardiography, and throughout the world where the expertise is often unavailable.

Other tools are under investigation for cardiology as well. Heart resonance MRI algorithms have been pursued by Arterys and Nvidia. These will hasten the interpretation of these scans, along with promoting accuracy.[31] But unlike ECGs and echo, MRI is not used frequently in the clinical setting. Besides images, even the conventional EHR has also been subject to machine algorithm scrutiny for predicting heart disease risk. The Nottingham group used EHR data from almost 380,000 patients, divided into a training cohort of more than 295,000 and validation in nearly 83,000 people.[32] Four different algorithms, including a neural network, exceeded the widely used American College of Cardiology/American Heart Association standard for ten-year predicted risk. The incorporation of socioeconomic class and ethnicity by the machine learning algorithms, in part, accounted for the clear advantage. Similarly, instead of the classic Framingham clinic risk factors that have been used to predict heart disease for several decades, a group at Boston University used machine algorithmic processing of EHR to achieve over 80 percent accuracy as compared with Framingham's 56 percent accuracy—scarcely better than a coin flip.[33]

CANCER DOCTORS

When IBM Watson first headed toward healthcare, it was little surprise that the field of cancer medicine was at the top of the list. With all the ways to define a person's cancer, there is probably no specialty in medicine that is as data rich and for which the extensive datasets are shaping state-of-the-art diagnosis and management. We know that each individual's cancer is unique and can be characterized in multiple layers. These include sequencing the person's native DNA, sequencing the tumor DNA, sequencing the tumor RNA, sequencing the circulating tumor DNA in the plasma (known as a liquid biopsy), characterizing the immune system status of both the tumor and the patient, and potentially growing the cancer cells in a dish to test the response of what is called an organoid to various drugs. The layers of information were recently extended to the analysis of live cancer cells, isolated with microfluidics from patients with either breast or prostate cancer, and evaluated by AI machine vision for predicting risk after surgery.[34] This is unique in the historical assessment of cancer, which until now relied on tissue blocks fixed in formalin (which certainly can be regarded as "dead"). Many of these biologic layers of data can or should be assessed serially: during treatment, during surveillance, or if there is recurrence. Add to that all the imaging information, and you've got terabytes of data about the person and his or her cancer process. Not only is there Big Data for each patient, there are more than 15 million Americans living with cancer, along with all their demographic, treatment, and outcome information.[35] To optimize outcomes, it is widely anticipated that combinations of therapies from different classes will be required, including, for example, treatments that are aimed at specific genomic mutations in the tumor and those that rev up the patient's immune system. The number of permutations and combinations are human-mind-boggling. There are plenty of reports now of successfully using two different immunotherapy treatments, and now that class of drugs has further expanded with the ability to engineer the patient's T cells. In short, the world of cancer is remarkably complex and represents a formidable challenge for expert clinicians, computational biologists, and artificial intelligence.

For breast cancer, I've reviewed the imaging and pathology slide AI efforts in Chapter 6. One study out of Methodist Hospital in Houston showed how AI significantly hastened mammography interpretation;[36] in another from Boston centers, machine learning of high-risk biopsy lesions predicted that surgery could have been avoided in 30 percent of more than a thousand patients.[37] But these studies are far from the challenge that lies ahead for AI to process inordinate datasets and improve clinical outcomes.

IBM Watson has had some false starts with MD Anderson, one of the leading cancer centers in the United States, but that was just one of more than fifty hospitals on five continents that have used Watson for Oncology.[38] Their collaboration with the UNC Lineberger Center turned out to be the source of the IBM Watson researchers' first peer-reviewed publication. A year after they presented their findings on *60 Minutes* in a segment titled "Artificial Intelligence Positioned to Be a Game-Changer,"[39] they provided the details of 1,018 patients with cancer, who had been previously reviewed by the UNC molecular tumor board, whose records were analyzed by IBM Watson.[40] The system found 323 patients that had what it termed actionable cancer, meaning they had tumor gene mutations that were suitable for drug trials, that the UNC team had overlooked. This automated analysis took less than three minutes per patient, which is astounding, but the IBM Watson team's conclusion was hyperbolic: "Molecular tumor boards empowered by cognitive computing could potentially improve patient care by providing a rapid, comprehensive approach for data analysis and consideration of up-to-date availability of clinical trials."[41] In fact, this really does not represent "cognitive computing," a term that IBM likes to use because, according to IBM's Ashok Kumar, it "goes beyond machine learning and deep learning."[42] I think it's funny, since Watson merely achieved an automated rather than manual curation, or matching patient mutations with clinical trials. There were no hidden layers, no deep learning. This was hardly a comprehensive approach. The results led Cory Doctorow, a thoughtful tech pundit, to conclude that "Watson for Oncology isn't an AI that fights cancer; it's an unproven 'mechanical turk.'"[43] Doctorow, for those not familiar with the original mechanical turk, was referring to a fake chess-playing machine—or as he elaborated, "a human-driven engine masquerading as an artificial intelligence"—that became notorious in the eighteenth century. Later, we learned that Watson for Oncology's "AI" guidance, which at times deviated from established guidelines and generated erroneous and even dangerous treatment recommendations, was based on the experience of a limited number of oncologists at Memorial Sloan Kettering.[44]

I was beginning to think we were a long way off. Then, serendipitously, I learned about Tempus Labs, a company started in 2015 by Eric Lefkofsky, the founder of Groupon, that is working on cancer. I wouldn't have thought to connect Groupon coupons with the future of cancer. But after his wife developed breast cancer in 2014, Lefkofsky found that there was no clinical or research entity positioned to make a difference. He said, "I was perplexed at how little data permeated care. It became apparent that the only way to usher in precision medicine was to fix the underlying data infrastructure in cancer."[45] Lefkofsky, a multibillionaire, decided to step in.

Lefkofsky has no science background, but you wouldn't know it if you spent just a little bit of time with him. I visited the company in the fall of 2017 and toured with Lefkofsky, coming away with the perspective that this was the first company taking a progressive and comprehensive approach to cancer. The facilities, housed along with Groupon in the 1.2 million square feet of what previously was the Montgomery Ward department store in downtown Chicago, were impressive in their own right. A large warehouse floor was filled with what seemed like an endless cohort of young scientists poring over data, such as unstructured doctor notes, at large monitors on their desks. Lefkofsky told me there were already more than a hundred employees with AI talent, and he was having no difficulty in attracting some of the best people in the country. During the tour, I saw the latest Illumina HiSeq and NovaSeq sequencing machines, a room for organoid cultures of cancer cells, another large area for machine learning of scans and biopsy reports, and an imaging room where pathology slides were magnified and projected bigger than life, with remarkable resolution. Here humans, seemingly, could diagnose pathology findings far better than if they were squinting into the lens of a microscope. At the time when I visited, Tempus had data from more than 11,000 patients and over 2.5 petabytes of data. With their cloud-based platform, cluster computing, natural-language processing, and AI capabilities, there was the infrastructure to build "the world's largest library of molecular and clinical data and an operating system to make that data accessible and useful."[46]

Tempus Labs, now collaborating with more than forty of the National Cancer Institute centers in the United States, performs a range of studies, including the list above from sequencing to culturing. Beyond the extensive assessment of the patient, Tempus provides "digital twin" information with their report generated two to three weeks after samples are received. This consists of treatment and outcomes information from the de-identified patients most similar with respect to demographics and biologic information. That, too, employs an AI advanced analytic method of nearest neighbor analysis.

Overall, it's a model based on deep phenotyping and deep analytics to help oncologists make data-driven decisions. Although it's too early to tell whether Tempus is the real deal, in two years it appears that the company has transcended what IBM Watson has done in more than five years with orders of magnitude more of capital and resources. When I asked Lefkofsky why more people in the medical community don't know about Tempus and why the company itself remains on the quiet side, he responded they had no interest in being the next Theranos. The company is intent on total transparency and publishing its data in peer-reviewed journals. That's laudable.*

Other companies besides IBM Watson and Tempus Labs are working to promote the use of AI, integrating multimodal data for cancer care. One is SOPHiA GENETICS, based in Switzerland, already being used in more than four hundred institutions in fifty-five countries. It brings together clinical, molecular, and imaging data to help guide oncologists.[47]

There is one last place so far where cancer could prove amenable to attack by AI: gastroenterology. The accurate diagnosis of colon polyps and cancer lesions during colonoscopy is more difficult than most people are aware.[48] Multiple studies have shown that these lesions are missed in at least 20 percent of patients, with some reports

---

* In 2018, after the book was written, I took on an advisory role for Tempus Labs to help expand their data-driven model to other diseases such as diabetes.

considerably higher. Lesions are more apt to be missed when they are flat, small, and in certain locations. Human eyes, even those of highly trained gastroenterologists, may not be as good as computer optical vision, as was suggested by a computer-aided study of more than two hundred small polyps.[49] Recently, the idea of using AI to detect these lesions was advanced in a deep learning study that used three hundred features from 30,000 colonoscopy images, magnified five-hundred-fold, and then tested the algorithm in 250 patients with 306 polyps.[50] The 86 percent accuracy achieved is promising compared with the literature. In the first prospective study of colonoscopy with real time AI-processed imaging, the results from 325 patients were very encouraging for accurate diagnosis of tiny (as they are called "diminutive") polyps.[51] The use of such high magnification and machine pattern review suggests it may ultimately be a very useful adjunct for this important cancer screening procedure.

SURGEONS

It might be counterintuitive to think that AI would have much of an influence on the hands and skills of surgeons. Performing operations is perhaps the furthest conceptually from straightforward input like a slide or scan. And ironic that surgery involves human touch even though the *Urban Dictionary* defines "human-touch" as "someone has empathy with other people's feelings and understanding of human side of things, not behave like a robot." For almost two decades, surgeons have been using robots, primarily the da Vinci of Intuitive Surgical, for AI-assisted operations. Although data from randomized trials of these robots have been unimpressive for improving key outcomes as compared with standard surgery,[52] in 2016 alone more than 4,000 such robots helped perform 750,000 operations around the world.[53] Still, that's less than 10 percent of the more than 8 million surgeries performed every year. Recent attempts to increase the prevalence and power of robotic surgery include Versius, a robot with arms that are much more like human arms, built by Cambridge Medical Robotics in the UK.[54] Other start-up companies with new robots include Medical Microinstruments, which has miniature wrists, without the need for a control console (well suited for microsurgery), and Auris Health, with FDA approval in 2018, whose robot functions as an endoscope.[55] It is placed into a patient's body via the mouth, to the trachea, and into the lung to obtain a tissue biopsy with computer-assisted vision. Medtronic acquired a German robot company that has haptic touch sensors, giving it more of a human surgeon sense of feel. There are already robots that can put in a row of sutures without human intervention and many contemplated applications for detection and debridement of dead or cancerous tissue. The recent progress of building robots with tactile sensation (unrelated to performing surgery) suggests we'll see further impact in operations going forward.[56] A small randomized trial of the first robotic-assisted microsurgery inside the human eye produced encouraging data for improving outcomes of this particularly delicate type of operation.[57]

While all these companies are improving upon robots that intrinsically require AI, Verb Surgical, which was formed in 2015 as a Google and Johnson & Johnson joint venture, is taking AI much further into the operating room. All Verb Surgical robots are connected to one another via the Internet, recording data of each procedure and applying machine learning to determine best surgical practice.[58] Calling it "Surgery 4.0," Verb's concept of cloud-connected surgeons sharing experiences and access to data is akin to democratizing surgical practice. In particular, machine learning that draws upon intraoperative imaging as well as all the relevant data from each patient could help redefine past practice and improve outcomes. For example, such an approach could identify the key surgical steps to avoid the serious and not uncommon complications of prostatectomy that include sexual dysfunction and urinary incontinence. There's also the integration of virtual reality and 3-D video microscopy to provide extraordinary visualization of anatomy during surgery.

So, however counterintuitive it might seem to imagine AI replacing surgeons, it's seeming increasingly possible. When a group of

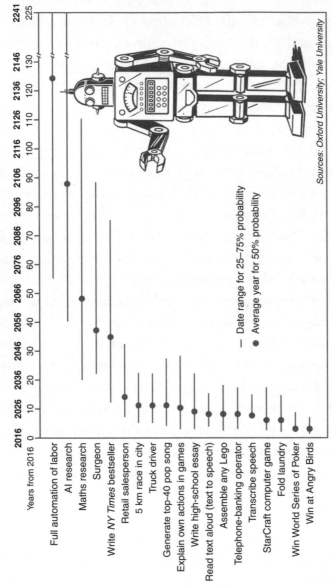

Man vs. Machine
Predicted year machines will match human performance

— Date range for 25–75% probability
● Average year for 50% probability

Sources: Oxford University; Yale University

FIGURE 7.3: Projections for various professions for when machines will match human performance. Note the projection for surgeons and AI research—and that this survey was conducted by computer scientists. Sources: Adapted from K. Grace et al., *When Will AI Exceed Human Performance? Evidence from AI Experts*, arXiv (2017): https://arxiv.org/abs/1705 .08807; *The World in 2017, Economist.*

researchers from Oxford and Yale Universities did a survey on when AI will exceed human performance in various fields (Figure 7.3), the consensus for replacing surgeons was approximately thirty years, twice as long as it would take to replace retail salespeople, but far less than the eighty-five years projected for AI researchers (as represented by the authors)![59]

OTHER HEALTHCARE PROFESSIONALS

Eventually, no type of clinician will be spared. We've already seen how neurologists can diagnose stroke more quickly with AI of brain imaging sent as texts to their smartphones.[60] A deep neural network for interpretation and urgent referral triage of more than 37,000 head CT scans demonstrated marked time-saving potential (150 times faster: 1.2 seconds for algorithm compared with 177 seconds for radiologists), but its accuracy was not acceptable (AUC = 0.56 for screening threshold).[61] We can't sacrifice accuracy for speed, but at least one significant component was improved upon with machine support. Like the Moorfields eye condition study that assessed patient OCT images for urgent referral, this study widened the use of a deep learning algorithm for help making the call for an urgent referral.[62] While deep learning AI is still taking on only narrow tasks, these two reports show widening beyond a single clinical diagnosis to suggesting an urgent referral for tens of potential diagnoses. We'll get into many other clinician types subsequently. One group that it is hard for me to envision ever being replaced in the deep medicine future is nurses, the real people caring for the sick. The robot nurse helper, Tug, at many hospitals, that delivers food and medications is certainly not a threat.[63] That isn't to say there are not opportunities for AI to augment nurses' work. For example, machine vision could keep track of patients in the intensive care unit to anticipate and prevent a patient pulling out an endotracheal tube, the uncomfortable but important device for helping patients breathe. Real-time analytics of vital signs, integrated with relevant lab and imaging data, could also help alert nurses to an impending problem. Other, more advanced robots and AI will be capable of taking and monitoring vital signs. And, as with physicians, there will be many tools to assist nurses as they increasingly have to deal with large datasets about their patients. But none of those tasks are the same as listening to, understanding, empathizing with, or simply holding hands with someone who is sick or just got news of a serious illness. I don't know that deep learning or robots will ever be capable of reproducing the essence of human-to-human support.

But AI could ultimately reduce the need for nurses, both in hospitals and in outpatient clinics and medical offices. Using AI algorithms to process data from the remote monitoring of patients at home will mean that there is a dramatically reduced role for hospitals to simply observe patients, either to collect data or to see whether symptoms get worse or reappear. That, in itself, has the potential for a major reduction in the hospital workforce. Increasing reliance on telemedicine rather than physical visits will have a similar effect.

The last in-depth look at how AI is changing specific fields will be in mental health. It's hard to think of the formidable challenge of digitizing a person's state of mind as anything that resembles a simple pattern. The topic is so big it requires a dedicated chapter. That's what's next.

chapter eight

MENTAL HEALTH

> After 40 years of psychiatry becoming more mindless
> than brainless, perhaps digital phenotyping will help the
> pendulum swing back toward a fresh look at behavior,
> cognition, and mood.
>
> —TOM INSEL

EVERY THURSDAY MORNING, I HAVE A RITUAL: I SCAN THE NEW edition of the *Economist*. Its science section usually covers three or four interesting topics that are not widely covered. One of the most memorable articles I've ever come across there had to do with humans preferring to trust machines with their innermost secrets instead of other humans—specifically, doctors. Indeed, the article referenced a paper I never would have seen in a journal I hadn't heard of—*Computers in Human Behavior*. The subtitle captured the point: "A Virtual Shrink May Sometimes Be Better Than the Real Thing."[1] That hadn't dawned on me before. But the research the *Economist* described clearly has profound implications in an age with an enormous mental health burden and limited professional personnel to provide support.

The study, led by Jonathan Gratch, was part of some innovative virtual human research at the Institute of Creative Technologies in

Los Angeles.[2] Gratch and his team had recruited 239 people from Craigslist. The only entry criteria were that participants be of ages eighteen to sixty-five and have good eyesight. All the study participants were interviewed by a human avatar named Ellie, which they saw via a TV screen. Half were randomly assigned to a group in which the experimenters told the participants that Ellie was not human, whereas the other half were told that Ellie was being controlled remotely by a person. The questions Ellie asked got progressively more intimate and sensitive, like "Tell me the last time you felt really happy."[3] The participants' faces were monitored, and the transcripts of the interviews were reviewed by three psychologists who did not know which participants were told Ellie was computer controlled and which were told she was human controlled. These data were used to quantify fear, sadness, and the participants' other emotional responses during the interviews, as well as their openness to the questions.

By every measure, participants were willing to disclose much more when they thought they were communicating with a virtual human rather than a real one. A couple of the participants who interacted with the virtual human conveyed this well: "This is way better than talking to a person. I don't really feel comfortable talking about personal stuff to other people." And "A human being would be judgmental. I shared a lot of personal things, and it was because of that."[4]

The findings provided strong empirical evidence for an idea that was first introduced in 1965, when people first bared their souls to ELIZA (named after Eliza Doolittle, a character in George Bernard Shaw's *Pygmalion,* and note the likeness to Ellie in this study), a very early computer program from MIT's Joseph Weizenbaum that was supposed to mimic a psychotherapy session, turning the person's answers into questions.[5] But evidence that this could work using a virtual human took many decades to be built. With Gratch's work, however, it seems that for deep thoughts to be disclosed, the avatars have a distinct advantage over humans. Indeed, at a 2018 *Wall Street Journal* health conference that I participated in, the majority of attendees polled said they'd be happy to, or even prefer to, share their secrets with a machine rather than a doctor. And, on a related note, an interesting Twitter poll with nearly 2,000 people responded to "You have an embarrassing medical condition. Would you rather tell and get treatment from (1) your doctor, (2) a doctor/nurse, or (3) a bot?" The bot narrowly beat "your doctor" by 44 percent to 42 percent.[6]

Although Gratch's study didn't seek out people with mental health problems, in recent years digital tools have been developed specifically for people having mental or emotional distress. Some of these have connected users with humans they don't know. Notably, 7 Cups of Tea (now simply called 7 Cups), which launched in 2013, provides free online chatting with volunteer trained listeners. By 2017, more than 230,000 listeners using 140 languages had helped more than 25 million people in 189 countries. About half were in the United States. Other examples include the Talkspace app, which has more than half a million users, and in the UK a National Health Service pilot study of a similar app has enlisted 1.2 million Londoners. Other tools have connected humans with chatbots using natural-language processing. In 2017, 8 million people in the United States talked to Cleverbot just to have something to chat with, and researchers project that by 2025 more than 1 billion people will be having regular encounters.[7] In China, Microsoft has released chat software called Xiaoice that has quickly amassed more than 20 million registered users. Recently, companies have begun developing chatbots for mental health support. A high-profile example, called Woebot, has Andrew Ng as its chairman. Woebot had more users in its first few months than would see a psychologist in a hundred years.[8]

Until recent years, our assessment of behavior, mood, and cognition was largely subjective, during brief episodic visits in a contrived, clinical environment. When this did occur, it usually was to respond to mental health difficulties rather than to prevent them. Table 8.1 lists some of the various and still proliferating ways we can

Speech	Prosody, volume, vowel space, word choice, length of phrases, coherence, sentiment
Voice	Valence, tone, pitch, intonation
Keyboard	Reaction time, attention, memory, cognition
Smartphone	Physical activity, movement, communication, sociality, social media, tweets, emoji, Instagram
Face	Emotion, tics, smiles and length, look at ground, eye movements, eye contact
Sensors	Heart rate, heart rate variability, galvanic skin response, skin temperature, blood pressure, breathing pattern, number of sighs, sleep, posture, gestures

TABLE 8.1: Digital phenotyping of mental state: the variety of metrics that can be used to digitize state of mind.

now collect objective data to deeply phenotype mood and mental health state. The term "digital phenotyping" conveys the point that each feature can be digitized and produce a variety of metrics. Most of these can be obtained passively via smartphones in a patient's real world. With the addition of connected sensors, many of the physiological parameters can be unobtrusively gathered, often on a continuous basis. This would mean a large body of data for each individual that AI could process. As Tom Insel, former head of the National Institute for Mental Health, said, "Could anyone have foreseen the revolution in natural-language processing and artificial intelligence that is allowing voice and speech, collected on a smartphone, to become a possible early warning sign of serious mental illness?"[9]

These metrics can be applied to a range of problems. Researchers at the University of Southern California developed software that was able to use seventy-four acoustic features, including voice quality, shimmer, pitch, volume, jitter, and prosody to predict marital discord as well as or better than therapists.[10] This same research group later compared expert manual coding of interviews with the acoustic data. Machine learning algorithms based on voice not only captured more relevant information than the experts but also predicted outcomes significantly better.[11]

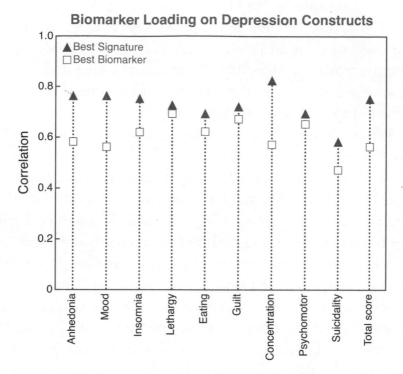

FIGURE 8.1: Correlation of biomarkers with feelings from Mindstrong using keyboard metrics, presented by Tom Insel, DigiMed Conference, La Jolla, California, October 5, 2017.

A small study of thirty-four youths, with an average age of twenty-two, undertook a "coherence" analysis of many features of speech such as length of phrases, muddling, confusion, and word choice to predict whether patients at risk of schizophrenia would transition to psychosis. The machine outperformed expert clinical ratings.[12] NeuroLex Diagnostics was formed to make a tool commercially available for primary-care doctors to diagnose schizophrenia, bipolar disorder, and depression and has developed a prototype that works on Amazon's Alexa.[13]

Even the way people use a smartphone's keyboard can be a useful marker. The company Mindstrong has broken this behavior down to forty-five patterns, including scrolling and latency time between space and character types. Their data correlated with gold-standard measurements for cognitive function and mood (Figure 8.1) in initial studies.

Computer scientists at the University of Illinois took this concept further with deep learning and a custom keyboard, loaded with an accelerometer. Using an algorithm they built called DeepMood, they predicted depression with very high accuracy in a pilot study, providing some independent proof of concept for passive mood tracking via an individual's keyboard activity.[14]

Some companies are already making inroads into clinical mental health practice with their tools. One, called Cogito, was cofounded by Alex "Sandy" Pentland, an accomplished polymath and professor at MIT whom I've met and greatly respect, and Joshua Feast. Pentland has been active in many areas of the digital revolution, notably preservation of privacy and security. (Only recently did I learn that he "learned to read in a mental institution" where his grandmother worked.) Sandy's Human Dynamics lab has studied "honest signals," the ways we unconsciously and nonverbally communicate truths about ourselves, for decades. Some examples of honest signals include our tone, fluidity, conversational engagement, and energy while we speak. Cogito used deep learning algorithms and honest signals to build an app called Companion that is used by psychologists, nurses, and social workers to monitor the mental health of their patients. By recording and uploading an audio diary entry, the app can assess the patients' status, picking up cues of depression and mood changes, by analyzing how they speak. It also can perform real-time analysis of conversation, which has been used by health insurance companies to handle customer calls.[15] The Companion app has also been used by the US Department of Veterans Affairs to monitor the mental health of at-risk veterans with 24/7 phone monitoring.[16]

Even Instagram photos are instructive. This social media platform is used far more than Twitter, with over 100 million new posts every day, growing faster than Facebook. In 2017, Andrew Reece and Christopher Danforth used deep learning to ingest 43,950 Instagram photos from 166 people (who digitally consented to share their social media history), of whom seventy-one had a history of depression.[17] As many photo features as you can imagine, and more,

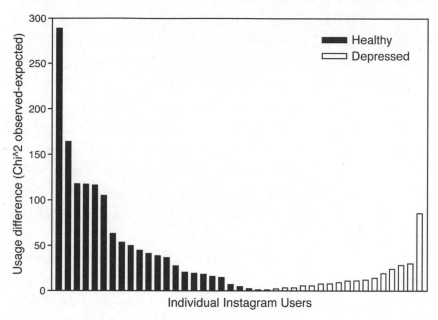

FIGURE 8.2: Instagram filter use for depressed and healthy individuals, with bars indicating difference between observed and expected frequencies. Source: Adapted from A. Reece and C. Danforth, "Instagram Photos Reveal Predictive Markers of Depression," *EPJ Data Science* 6 (2017): 15.

were analyzed for psychological insight: whether people were present; whether the setting was indoors or outdoors, night or day; color and brightness by pixels; comments and likes of the photos; and the posting frequency of the user. The Instagram photos differentiated depressed and healthy individuals, could be used to diagnose depression before it was clinically diagnosed, and did not correlate with the person's self-assessment of mental health. Notably, such features as the Instagram filter to remove color partitioned depression and healthy people more than anticipated (Figure 8.2). The machine accuracy for detection of depression was 70 percent, which compared favorably to general practice doctors, as previously published, who have a false positive diagnosis of depression of over 50 percent.[18] Psychiatrists are better, but the vast majority of people with depression are seen by a primary-care doctor or aren't seen by any clinician at all, let alone a psychiatrist.

Approaches like these are being harnessed to explore new means to diagnose and treat a range of psychiatric and emotional problems, but it's important to keep in mind a few caveats. For one, it's noteworthy that many of the sensors have not been validated for accuracy, and they aren't necessarily measuring what they purport to measure. For example, sleep quality—an important metric for a range of mental health issues—is often assessed through a wristband or watch that simply senses the wearer's movements during sleep. But to truly know the patient's sleep state, those movements would need to have established correlations with brain waves, which has not been done.

The biomarkers may also be too simple. As Nunzio Pomara, a professor at NYU said, "Depression is too complex to be reduced to a single biomarker."[19] We have so many biomarkers for mental health (Table 8.1), and we have no idea which ones, or how many of these, are critical to making an accurate diagnosis or for monitoring a treatment response. With seventy-four voice subfeatures and forty-five keyboard interaction subfeatures, and so on, we would need a computer just to figure out the millions of permutations and combinations. The studies performed to date—those I covered here and more that I haven't—have been small and very narrow, typically zooming in on a particular marker. It is likely that some combination of markers could be quite useful, but at this point we have no idea what the right combination is or whether it varies for each individual or condition: what works for post-traumatic stress disorder (PTSD) may be very inaccurate for depression. We also have no idea of what it takes to achieve saturation, when adding more markers fails to add to accuracy. With respect to establishing accuracy, the ground truths are tricky because historically mental health disorders have largely been defined by subjective and clinical features. The pragmatic aspects of how to collect data unobtrusively with inexpensive software-generated feedback have yet to be addressed. Having pointed out all these holes, I do nevertheless think that there's hope we'll get there someday, and, for those eager to see substantive progress, I think there's no reason to be depressed. But now let's tackle what we know about depression.

Depression is by far the most common mental health disorder, with more than 350 million people battling it every day.[20] Depression accounts for more than 10 percent of the total global burden of disease; worldwide more than 76 million human-years are lost to disability each year, which far outstrips heart disease, cancer, and all other medical diagnoses.[21] Each year 7 percent of Americans (16 million adults) will be clinically diagnosed with depression, and the lifetime risk of a mental disorder is approximately 30 percent. Of more than $200 billion per year the United States spends on mental health, the vast majority is tied to depression—and even with that great expenditure, not everyone is seen by a physician, let alone helped. In 2016, of the more than 16 million adults in the United States who had a major bout of depression, 37 percent received no treatment.[22] There is much room for improvement.

Until the biomarker era, depression was diagnosed by the *Diagnostic and Statistical Manual of Mental Disorders* (*DSM*) when a patient met five of nine criteria, including depressed mood, changes in sleep or physical activity, feelings of worthlessness, and decreased pleasure (anhedonia). Many of these are hard to assess quantitatively or objectively.

Several approaches have been attempted to make the diagnosis more quantitative. One approach involves measuring brain waves, which have been suggested as a way to diagnose several other mental health problems as well. Despite the problem that wearing head gear to provide brain electrical activity sure doesn't seem like a scalable or practical way to track state of mind, some employers in China are requiring their workers to wear caps for brain-wave monitoring.[23] There are no data that such caps capture high-fidelity brain waves, let alone whether those waves accurately determine one's emotional state. One thing it does demonstrate is the flagrant lack of regard for employees' privacy. Still, in the long term, be it by an unobtrusive wearable device or a chip implanted in the brain, it is theoretically possible (yet not at all alluring) that brain-wave data might be useful.

As a research tool, brain magnetic resonance imaging has been shown to be a powerful biomarker to characterize depression. Using diffusion tensor MRI measures of brain white matter and machine learning, major depression disorder was shown to be quite distinct from healthy controls.[24] Conor Liston and colleagues at Weill Cornell Medicine analyzed scans from nearly 1,200 people, of whom 40 percent were diagnosed with depression.[25] When the MRIs were subject to machine learning of the signal fluctuations from 258 brain regions, four distinct biotypes were identified (Figure 8.3). All four of these brain patterns of connectivity were different from healthy controls, and each had an associated symptom complex such as fatigue, low energy, insomnia, or anhedonia. The patterns also predicted treatment response for patients who underwent transcranial magnetic stimulation, which helped people with biotypes 1 and 3 (~70 percent effective) compared with biotypes 2 and 4 (25 percent response). When MRIs from patients with schizophrenia or generalized anxiety disorder (GAD) were compared, there was little overlap for the former, but most of the individuals with GAD fit into one of the four depression biotypes.

Similarly, machine learning algorithms have been used in other small studies with functional MRI brain images to identify patterns tied to major depression disorder as compared with healthy controls.[26]

Beyond the keyboard and Instagram studies, several ongoing studies are focusing on more everyday biomarkers such as voice and speech to diagnose and characterize depression, including Sonde Health's postpartum depression project[27] and New York University's Charles Marmar for PTSD.[28] Using neural networks, Marmar has identified thirty voice features that may differentiate veterans with PTSD compared with unaffected veterans or healthy controls, and

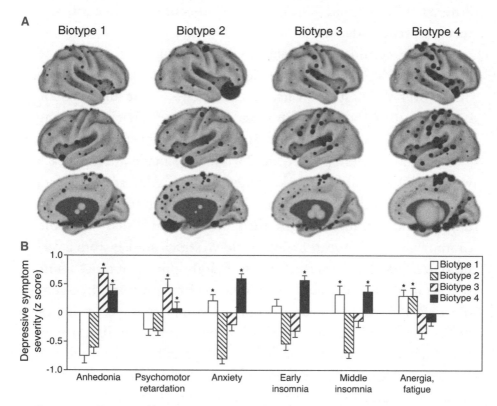

FIGURE 8.3: Functional brain connectivity markers for diagnosing neurophysiological bio-types of depression, correlating MRI brain signals (A) and patient symptoms (B). Source: Adapted from A. Drysdale et al., "Resting-State Connectivity Biomarkers Define Neuro-physiological Subtypes of Depression," *Nat Med* (2017): 23, 28–38.

these are being used in a five-year prospective large-cohort study of outcomes. Reduced vowel space has been shown by machine learning algorithms to be significantly altered in more than 250 individuals with PTSD.[29] Use of voice data for bipolar disorder in seven smartphone-based studies was summarized well in the review article's title: "High Potential but Limited Evidence."[30] Whether patients will respond to medications for depression is an area of great uncertainty in mental health, both because the effectiveness of the drugs is spotty and because there are so many medications to choose from. Machine learning algorithms have been used with clinical features to predict response to antidepression medications, but so far the accuracy has hovered at 60 percent, which is not too encouraging (like rolling a six on a twenty-sided die).[31]

There is considerable recent interest in using AI to predict and prevent suicide. The suicide rate has been increasing in the United States over the past 30 years, accounting for more than 44,000 deaths in 2017,[32] or over 120 suicides per day.[33] That's more than homicide, AIDS, car accidents, and war. The bigger picture from global data is striking: there are 25 million suicide attempts each year, and 140 million people contemplate it. Almost 80 percent of those who kill themselves hid their suicidal ideation from their doctors and therapists during their most recent visits.[34] A mammoth review of fifty years of 365 suicide research studies from 2,542 unique papers looking at more than 3,400 different metrics found that, at best, those thousands of risk factors are very weak predictors of suicidal ideation, suicide attempts, or completion—only slightly better than random guessing.[35] With no category or subcategory accurately predicting above chance levels, Joseph Franklin and colleagues concluded, "These findings suggest the need for a shift in focus from risk *factors* to machine learning algorithms."[36]

In 2017, a team of researchers at Vanderbilt and Florida State Universities did just that. After reviewing 2 million de-identified electronic medical records from Tennessee hospitalized patients, the researchers found more than 3,000 patients with suicide attempts. Applying an unsupervised learning algorithm to the data accurately predicted suicide attempts nearly 80 percent of the time (up to a six-month window), which compares quite favorably to the 60 percent from logistic regression of traditional risk factors.[37] The researchers pointed out that the algorithm could be improved if they had access to information such as life events, like marriage breakup or loss of a job, abrupt changes in mood or behavior, and social media data.[38]

Other studies have investigated those sorts of data. A machine learning algorithm developed at Cincinnati Children's Hospital by John Pestian was reported, in 479 patients, to achieve 93 percent accuracy for predicting serious risk of suicide.[39] It incorporated data on real-world interactions such as laughter, sighing, and expression of anger. Researchers at Carnegie Mellon did a very small but provocative study with functional MRI brain images of seventeen suicidal ideators and seventeen controls.[40] Machine learning algorithms could accurately detect "neurosemantic" signatures associated with suicide attempts. Each individual, while undergoing the MRI, was presented with three sets of ten words (like "death" or "gloom"). Six words and five brain locations determined a differentiating pattern. Machine learning classified the brain image response correctly in fifteen of the seventeen patients in the suicide group and sixteen of the seventeen healthy controls. This study is interesting from an academic perspective but of limited practical use because it is not likely we will ever be doing MRI scans to find people with suicide risk.

Researchers have also taken advantage of social media to identify suicide risk and emotional distress. Texts from the very popular Chinese platform Weibo were analyzed with machine learning with some word classifiers detected.[41] At a much larger scale, Facebook is mining its wall posts of users who report risk of self-harm. After Facebook Live was rolled out in 2016, it has been used by several people to broadcast their suicide. With a heightened awareness of the opportunity to prevent such tragedies, in 2017, CEO Mark Zuckerberg announced new algorithms that look for patterns of posts and words for rapid review by dedicated Facebook employees: "In the future, AI will be able to understand more of the subtle nuances of language and will be able to identify different issues beyond suicide as well, including quickly spotting more kinds of bullying and hate." Unfortunately, and critically, Facebook has refused to disclose the algorithmic details, but the company claims to have interceded with more than a hundred people intending to commit self-harm.[42]

Data scientists are now using machine learning on the Crisis Text Line's 75 million texts[43] to try to unravel text or emoji risk factors.[44] Overall, even these very early attempts at using AI to detect depression and suicidal risk show some promising signs that we can do far better than the traditional subjective and clinical risk factors. What is particularly interesting is that the technology may evolve into a closed loop, whereby devices such as a smartphone could be used to not only facilitate diagnosis but also serve as the conduit for therapy.

One approach is to bring cognitive behavioral therapy (CBT), a form of psychotherapy that has traditionally relied on intensive in-person sessions, to our phones. CBT has many definitions. Chief among these is to change patterns of maladaptive thinking or behavior—to "help people to identify and change negative, self-destructive thought patterns."[45] The digital version of CBT is more simply defined: talk therapy. It appears to have similar efficacy for treating depression (at least mild to moderate types) as the labor-intensive face-to-face visits with a mental health professional. There are plenty of CBT mobile apps including Lantern, Joyable, MoodGYM, and Ginger.io. A meta-analysis of eighteen randomized control trials of more than 3,400 patients using twenty-two smartphone apps for treating depression showed significant improvement, and those apps based on CBT were particularly effective.[46]

All those apps studied involve interactions with human beings, but not all apps rely on human interaction. Wysa, a penguin chatbot, has attracted 50,000 users who engaged in a million conversations in just three months; more than five hundred wrote in comments to say how much it helped with their mental health problem.[47] Woebot uses an instant messenger app to conduct sessions with users. It typically starts with open-ended questions like "What's going on in your world right now?" and "How are you feeling?" Natural-language processing both quantifies the user's state of mental health and defines the path of conversation to deliver CBT in response. Built by Alison Darcy, formerly on the faculty at Stanford University, this text-based, conversational, mood-tracking agent was put to the test in a small randomized trial of seventy college students assigned CBT alone or with Woebot.[48] There was better engagement with Woebot and signs of more reduced depression compared with the CBT strategy. This virtual counselor[49] was reviewed as clever or cute by some and as having "a personality that's a cross between Kermit the Frog and Spock from Star Trek" by others.[50] Another chatbot designed for psychotherapy is X2AI; it detects extensive data on speech phrasing, diction, typing speed, and grammatical voice to correlate with emotional state.[51]

If more studies build on the initial evidence and demonstrate improved outcomes at scale, the CBT and chatbot offerings may do well in the mental health arena, for which there is a woeful shortage of health professionals. In the United States and high-income countries, more than half of people with mental disorders do not receive care; that proportion goes up to 85 percent in low- and middle-income countries.[52] More than 106 million people in the United States live in areas that are designated by the federal government as short of mental health professionals.[53] The number of psychiatrists in the United States is less than 8 per 100,000 people, but in most low- to middle-income countries that number is far less than 1 per 100,000 (worst case, Afghanistan = 0.16).[54] So, although virtual counselors will never fully replace real human ones, this could turn out to be one of the most important booster functions of AI in medicine. It's software, it's cheap, it should just get better and better via deep learning. As Nick Romeo pointed out, "AI counselors need no plane tickets, food, protection, or salaries. They can easily handle caseloads in the tens of thousands, and they can be available at any time, via text message, to anyone with a mobile phone."[55] And that especially applies to young people, who are tightly connected to their smartphones and a principal group of concern: 74 percent of mental health disorders have their onset before the age of twenty-four.[56]

One of the big trepidations about chatbots and smartphone apps for mental health is about privacy and security of the data. In fact, it's quite remarkable how broadly mental health apps are used now in light of this issue going largely unaddressed. No matter how much progress has been made, mental disorders are still stigmatized, and it's such highly sensitive information that most people are deeply concerned about breach of their data or loss of privacy. Woebot and

Facebook have said they do not see any data or sell information or ads based on user content. But concerns over privacy, including hacking of psychological history and data, whether for sale or theft, remain an unsettled apprehension.

There is also the ethical concern about people talking to machines, sharing their intimate experiences and emotions. Writing in the *New Yorker*, Allison Pugh likened this to the "cloth monkey" experiments conducted in 1959 by the psychologist Harry Harlow. Harlow required the monkeys to make a cruel choice: either a surrogate mother made of cloth or one made of wire that also provided milk.[57] The dilemma symbolizes important choices for humans and their machines. The willingness of vulnerable people to relay their shame to machines may, in effect, preempt human caring. Some might argue that this form of AI treatment could ultimately prove to be worse than no therapy.

Culminating our discussion about the potential of AI to influence mental health by treating disease, I want to consider its potential for increasing happiness. Yuval Noah Harari suggests in his book *Homo Deus* that ensuring global happiness will be one of the three most important goals of what he calls the humanist revolution (along with maximizing human life span and power)—arguing even that we'll measure the civilizations of the future not by gross domestic product but gross domestic happiness.[58] Indeed, he claims that people really don't want to produce, they just want to be happy, and our technology and knowledge will be so remarkably advanced that we could distill the elixir of true happiness (and no god will stop us).

We're clearly a long, long way from the world Harari describes, if we can even get there. But, like depression, happiness is something we could potentially use technology to measure and improve. Pascal Budner and colleagues at MIT collected around 17,000 pieces of data, including heart rate, location, and weather conditions via smartwatches in a small study of sixty people over two months. The users entered their state-of-mind data four times a day using a "happimeter," picking from nine emojis.[59] While it's hard to conclude much from the study, it represents one of the first to use AI to understand and track the flip side of depression. Indeed, we are still at the earliest point of defining happiness; we do know what the principal reason for its absence is—mental health disorders. The *World Happiness Report* 2017 looked at all the known factors, including poverty, education, employment, partnership, physical illness, and mental illness. For the four countries it scrutinized, the United States, Australia, United Kingdom, and Indonesia, mental illness was clearly the dominant factor tied to misery, detracting from happiness.[60]

I think it's fair to say we'll never achieve world happiness, but the goal of alleviating the profound global burden of depression needs to be front and center. The biomarker revolution of digitizing depression coupled with the tools of AI have put us in good stead to make a serious dent in this problem.

Historically, we've prioritized physical over mental illness. It is easier to measure, easier to treat, and less stigmatized. But now we could be looking at a revolution in mental health, with more openness; new, objective biomarkers that afford the "digitization" of the mind; and therapies that don't fully rely on trained humans. In the face of a global mental health crisis with increasing suicides and an enormous burden of depression and untreated psychiatric illness, AI could help provide a remedy. Digital phenotyping of behavior and state of mind also has implications well beyond diagnosing mental illness. Social and behavioral dynamics are key to providing a holistic, multimodal picture—including physiological, biological, anatomical, and environmental data—for each individual. The seamless capturing and processing of such data may prove to be helpful in understanding the relationship of stress to common medical conditions like high blood pressure and diabetes.

However those different factors come together in each particular case, for each particular patient, in one way or another, healthcare professionals across the board will be influenced by AI. If radiologists perform gatekeeping functions limiting unnecessary scans, we may ultimately need fewer technicians to perform medical scans. When data from medical records, genomic screens, and sensors are integrated and processed by AI, pharmacists will be able to offer better guidance about prescriptions, such as suggesting a drug's lack of efficacy, an unanticipated but likely interaction, or a serious side effect. Physical therapists, as I learned firsthand, might be able to deliver more tailored programs that line up with each individual patient's detailed profile. AI generative adversarial networks can make dental crowns for individual patients more accurately than human experts can, facilitating restorations in dentistry.[61] Paramedics in Copenhagen have already been helped by an AI-assistant made by Corti, a company that uses speech recognition and data from emergency calls to accurately diagnose heart attacks.[62] With advanced analytics and output that include recommendations for each patient, AI is likely to empower nurse clinicians and physician assistants to take on more responsibility. As we think about each and every type of clinician, it becomes increasingly clear that AI has a potential transformative impact. But it's not just singularly at the level of clinicians; it's also at the level of the sum of the parts—the people—that make up a health system.

Now it's time to move beyond how AI might directly benefit patients and clinicians. We're ready to address how these tools can change the face of health systems, the subject of the next chapter.

chapter nine

AI AND HEALTH SYSTEMS

> The nurses wear scrubs, but the scrubs are very,
> very clean. The patients are elsewhere.
>
> —ARTHUR ALLEN

A FEW YEARS AGO, ON A WARM SUNNY AFTERNOON, MY ninety-year-old father-in-law was sweeping his patio when he suddenly felt weak and dizzy. Falling to his knees, he crawled inside his condo and onto the couch. He was shaking but not confused when my wife, Susan, came over minutes later, since we lived just a block away. She texted me at work, where I was just finishing my clinic, and asked me to come over.

When I got there, he was weak and couldn't stand up on his own, and it was unclear what had caused this spell. A rudimentary neuro exam didn't show anything: his speech and vision were fine; muscle and sensory functions were all okay save for some muscle trembling. A smartphone cardiogram and echo were both normal. Even though I knew it wouldn't go over too well, I suggested we take him to the emergency room to find out what the problem was.

John, a Purple Heart–decorated World War II vet, had never been sick. In fact, we had previously enrolled him in our Wellderly genomics sequencing research program at Scripps Research for people age eighty-five and older who had a remarkable health span, never had a chronic illness, and weren't taking medications like statins for high cholesterol or other chronic conditions. Only in recent months had he developed some mild high blood pressure, for which his internist had prescribed chlorthalidone, a weak diuretic. Otherwise, his only medicine over the years was a preventive baby aspirin every day.

With some convincing he agreed to be seen, so along with his wife and mine, we drove over to the local ER. The doctor there thought he might have had some kind of stroke, but a head CT didn't show any abnormality. But then the bloodwork came back and showed, surprisingly, a critically low potassium level of 1.9 mEq/L—one of the lowest I've seen. It didn't seem that the diuretic alone could be the culprit. Nevertheless, John was admitted overnight just to get his potassium level restored by intravenous and oral supplement.

All was well until a couple of weeks later, when he suddenly started vomiting bright red blood. He was so unwilling to be sick that he told his wife not to call Susan. But she was panicked and called Susan anyway. Again, my wife quickly arrived on the scene. There was blood everywhere, in the bedroom, in the living room, and bathroom. Her father was fully alert despite the vomiting and a black, tarry stool, both of which were clear indications that he was having a major gastrointestinal bleed. He needed to go to the ER again. At the hospital a few hours later, after an evaluation and a consultation with a GI specialist, an urgent endoscopy showed my father-in-law had esophageal varices that were responsible for the bleeding.

To do the procedure of localizing the source of bleeding, John was anesthetized and given fentanyl, and when he finally got to a hospital room in the evening, he could barely say a few words. Soon thereafter he went into a deep coma. Meanwhile his labs came back: his liver function tests were markedly abnormal, and his blood ammonia level was extremely high. An ultrasound showed a cirrhotic liver. We quickly came to the realization that the esophageal varices were secondary to end-stage liver disease. A man who had been perfectly healthy for ninety years all of a sudden was in a coma with a rotted liver. He was receiving no intravenous or nutritional support, but he was receiving lactulose enemas to reduce his blood ammonia level from the liver failure. His prognosis for any meaningful recovery was nil, and the attending doctor and the medical residents suggested that we make him a do-not-resuscitate order.

Arrangements were made over the next few days for him to come to our house with hospice support, so he could die at home. Late on a Sunday night, the night before we were to take my father-in-law home to die, my wife and daughter went to visit him. They both had been taught "healing touch" and, as an expression of their deep love, spent a few hours talking to him and administering this spiritual treatment as he lay comatose.

On Monday morning, my wife met with the hospice nurse outside the hospital room. Susan told the nurse that, before they went over the details, she wanted to go see her father. As Susan hugged him and said, "Dad, if you can hear me, we're taking you home today." John's chest heaved; he opened his eyes, looked at her, and exclaimed, "Ohhhhhhh." She asked him if he knew who she was, and he said, "Sue."

If there was ever a family Lazarus story, this was it. Everything was turned upside down. The plan to let him die was abandoned. When the hospice transport crew arrived, they were told the transfer plan was ditched. An IV was inserted for the first time. The rest of the family from the East Coast was alerted of his shocking conversion from death to life so that they could come to visit. The next day my wife even got a call on her cell phone from her father asking her to bring him something to eat.

My lasting memory of that time is taking John on a wheelchair ride outside. By then he'd been in the hospital for ten days and, now attached to multiple IVs and an indwelling Foley catheter, was as pale as the sheets. Against the wishes of his nurses, I packaged him up and took him in front of the hospital on a beautiful fall afternoon. We trekked down the sidewalk and up a little hill in front of the hospital; the wind brought out the wonderful aroma of the nearby eucalyptus trees. We were talking, and we both started to cry. I think for him it was about the joy of being alive to see his family. John had been my adopted father for the past twenty years, since my father had died, and we'd been very close throughout the nearly forty years we had known each other. I never imagined seeing him sick, since he had always been a rock, the veritable picture of health. And now that he had come back to life, compos mentis, I wondered how long this would last. The end-stage liver disease didn't make sense, since his drinking history was moderate at worst. There was a blood test that came back with antibodies to suggest the remote possibility of primary biliary cirrhosis, a rare disease that didn't make a lot of sense to find in a now-ninety-one-year-old man (the entire family had gotten to celebrate his birthday with him in the hospital). Uncertainties abounded.

He didn't live much longer. There was debate about going to inject and sclerose the esophageal varices to avoid a recurrent bleed, but that would require another endoscopy procedure, which nearly did him in. He was about to be discharged a week later when he did have another bleeding event and succumbed.

PREDICT, PREDICT, PREDICT

What does this have to do with deep changes with AI? My father-in-law's story intersects with several issues in healthcare, all of them centering on how hospitals and patients interact.

The most obvious is how we handle the end of life. Palliative care as a field in medicine is undergoing explosive growth already. It is going to be radically reshaped: new tools are in development using the data in electronic health records to predict time to death with unprecedented accuracy while providing the doctor with a report that details the factors that led to the prediction.[1] If further validated, this and related deep learning efforts may have an influence for palliative care teams in more than 1,700 American hospitals, about 60 percent of the total. There are only 6,600 board certified palliative-care physicians in the United States, or only 1 for every 1,200 people under care, a situation that calls out for much higher efficiency without compromising care. Less than half of the patients admitted to hospitals needing palliative care actually receive it.[2] Meanwhile, of the Americans facing end-of-life care, 80 percent would prefer to die at home, but only a small fraction get to do so—60 percent die in the hospital.[3]

A first issue is predicting when someone might die—getting that right is critical to whether someone who wants to die at home actually can. Doctors have had a notoriously difficult time predicting the timing of death. Over the years, a screening tool called the Surprise Question has been used by doctors and nurses to identify people nearing the end of life—to use it, they reflect on their patient, asking themselves, "Would I be surprised if this patient died in the next twelve months?" A systematic review of twenty-six papers with predictions for over 25,000 people, showed the overall accuracy was less than 75 percent, with remarkable heterogeneity.[4]

Anand Avati, a computer scientist at Stanford, along with his team, published a deep learning algorithm based on the electronic health record to predict the timing of death. This might not have been clear from the paper's title, "Improving Palliative Care with Deep Learning," but make no mistake, this was a dying algorithm.[5] There was a lot of angst about "death panels" when Sarah Palin first used the term in 2009 in a debate about federal health legislation, but that was involving doctors. Now we're talking about machines. An eighteen-layer DNN learning from the electronic health records of almost 160,000 patients was able to predict the time until death on a test population of 40,000 patient records, with remarkable accuracy. The algorithm picked up predictive features that doctors wouldn't, including the number of scans, particularly of the spine or the urinary system, which turned out to be as statistically powerful, in terms of probability, as the person's age. The results were quite powerful: more than 90 percent of people predicted to die in the following three to twelve months did so, as was the case for the people predicted to live more than twelve months. Noteworthy, the ground truths used for the algorithm were the ultimate hard data—the actual timing of deaths for the 200,000 patients assessed. And this was accomplished with just the structured data in the electronic records, such as age, what procedures and scans were done, and length of hospitalization. The algorithm did not use the results of lab assays, pathology reports, or scan results, not to mention more holistic descriptors of individual patients, including psychological status, will to live, gait, hand strength, or many other parameters that have been associated with life span. Imagine the increase in accuracy if they had—it would have been taken up several notches.

An AI dying algorithm portends major changes for the field of palliative care, and there are companies pursuing this goal of predicting the timing of mortality, like CareSkore, but predicting whether someone will die while in a hospital is just one dimension of what neural networks can predict from the data in a health system's electronic records.[6] A team at Google, in collaboration with three academic medical centers, used input from more than 216,000 hospitalizations of 114,000 patients and nearly 47 billion data points to do a lot of DNN predicting: whether a patient would die, length of stay, unexpected hospital readmission, and final discharge diagnoses were all predicted with a range of accuracy that was good and quite consistent among the hospitals that were studied.[7] A German group used deep learning in more than 44,000 patients to predict hospital death, kidney failure, and bleeding complications after surgery with remarkable accuracy.[8] DeepMind AI is working with the US Department of Veterans Affairs to predict medical outcomes of over 700,000 veterans.[9] AI has also been used to predict whether a patient will survive after a heart transplant[10] and to facilitate a genetic diagnosis by combining electronic health records and sequence data.[11] Mathematical modeling and logistic regression have been applied to such outcome data in the past, of course, but the use of machine and deep learning, along with much larger datasets, has led to improved accuracy.

The implications are broad. As Siddhartha Mukherjee reflected, "I cannot shake some inherent discomfort with the thought that an algorithm might understand patterns of mortality better than most humans."[12] Clearly, algorithms can help patients and their doctors make decisions about the course of care both in palliative situations and those where recovery is the goal. They can influence resource utilization for health systems, such as intensive care units, resuscitation, or ventilators. Likewise, the use of such prediction data by health insurance companies for reimbursement hangs out there as a looming concern.[13]

Hospitals are the number one line item of healthcare costs in the United States and account for almost one-third of the $3.5 trillion annual expenditures. While personnel are the biggest factor driving their costs, the need to avoid hospitalizations, whether the first or a readmission, has taken center stage for many AI initiatives. Economics play a major role here, since readmission within thirty days of hospitalization may not be reimbursable. There are concerns, and indeed some debate, as to whether trying to restrict hospitalizations would have an adverse effect on patient outcomes.[14]

Multiple studies have taken on the challenge of predicting whether a hospitalized patient will need to be readmitted in the month following discharge from a hospital, particularly finding features that are not captured by doctors. For example, a study conducted by Mount Sinai in New York City used electronic health records, medications, labs, procedures, and vital signs, and demonstrated 83 percent accuracy in a relatively small cohort.[15] A much larger set of 300,000 patients was used to train and validate the DeepR Analytics DNN,[16] which compared favorably to other efforts, including DoctorAI[17] and DeepCare. This objective is being pursued by many start-up companies and academic centers, along with AI for case management. Notably, Intermountain Healthcare,

University of Pittsburgh Medical Center, and Sutter Health are among first movers working on implementation of such algorithms.

Among the bolder objectives is to predict disease in patients without any classic symptoms. A group based at Tsinghua University in Beijing took data from more than 18,000 real-world EHRs to accurately diagnose six common diseases: hypertension, diabetes, COPD, arrhythmia, asthma, and gastritis.[18] Using solely eighteen lab tests, certain conditions, such as kidney disease, could be accurately predicted by a DNN in a large cohort of nearly 300,000 patients followed over eight years.[19] The Mount Sinai group studied EHRs from 1.3 million patients to predict five diseases—diabetes, dementia, herpes zoster (also known as shingles), sickle cell anemia, and attention deficit disorder—with very high accuracy. Preventing these diseases would require two variables to be true: that such algorithms using EHRs, lab tests, and other data survive further testing, showing that they can indeed predict the onset of these diseases, and that effective treatments are available. If both are true, the algorithms could become useful not just to reduce the human burden of disease but also to aid employers and insurers in their quest to reduce costs. However, at this juncture, all the prediction has taken place in silico—from preexisting datasets in machines, not in a real-world clinical environment. For a sample of fifteen studies looking at predicting a variety of outcomes (Table 9.1), there were significant statistical methodologic deficiencies in most, along with considerable variability in sample size and level of accuracy.[20] We simply don't know yet how well AI will be able to predict clinical outcomes.

Going back to my father-in-law's case, his severe liver disease, which was completely missed, might have been predicted by his lab tests, performed during his first hospitalization, which showed a critically low potassium level. AI algorithms might have even been able to identify the underlying cause, which remains elusive to this day. My father-in-law's end-of-life story also brings up many elements that will never be captured by an algorithm. Based on his

Prediction	N	AUC	Reference
In-hospital mortality, unplanned readmission, prolonged LOS, final discharge diagnosis	216,221	0.93* 0.75+ 0.85#	Rajkomar et al., Nature NPJ Digital Medicine, 2018
All-cause 3-12 month mortality	221,284	0.93^	Avati et al., arXiv, 2017
Readmission	1,068	0.78	Shameer et al., Pacific Symposium on Biocomputing, 2017
Sepsis	230,936	0.67	Horng et al., PLOS One, 2017
Septic shock	16,234	0.83	Henry et al., Science, 2015
Severe sepsis	203,000	0.85@	Culliton et al., arXiv, 2017
C. difficile infection	256,732	0.82++	Oh et al., Infection Control and Epidemiology, 2018
Developing diseases	704,587	range	Miotto et al., Scientific Reports, 2018
Diagnosis	18,590	0.96	Yang et al., Scientific Reports, 2018
Dementia	76,367	0.91	Cleret de Langavant et al., J Internet Med Res, 2018
Alzheimer's disease (+ amyloid imaging)	273	0.91	Mathotaarachchi et al., Neurobiology of Aging, 2017
Mortality after cancer chemotherapy	26,946	0.94	Elfiky et al., JAMA Open, 2018
Disease onset for 133 conditions	298,000	range	Razavian et al., arXiv, 2016
Suicide	5,543	0.84	Walsh et al., Clinical Psychological Science, 2017

AUC: area under the curve, a metric of accuracy, LOS: length of stay, N: number of patients (training + validation datasets), *in-hospital mortality, +unplanned readmission, #prolonged LOS, ^all patients, @structured and unstructured data, ++for U of Michigan site

TABLE 9.1: A sample of fifteen studies using AI to predict clinical outcomes.

labs, liver failure, age, and unresponsiveness, his doctors said he would never wake up and was likely to die within a few days. A predictive algorithm would have ultimately been correct that my father-in-law would not survive his hospital stay. But that doesn't tell us everything about what we should do during the time in which my father-in-law, or any patient, would still live. When we think of human life-and-death matters, it is hard to interject machines and algorithms—indeed, it is not enough. Despite the doctors' prediction, he came back to life and was able to celebrate his birthday with his extended family, sharing reminiscences, laughter, and affection. I have no idea whether human healing touch was a feature in his resurrection, but my wife and daughter certainly have their views on its effect. But abandoning any efforts to sustain his life would have preempted the chance for him to see, say good-bye to, and express his deep love for his family. We don't have an algorithm to say whether that's meaningful.

THE HEALTHCARE WORKFORCE AND WORKFLOW

There are many uses of AI for hospitals and health systems that extend well beyond predicting death and major outcomes. By 2017, healthcare became the number one US industry by total jobs for the first time, rising above retail.[21] More than 16 million people are employed by health services, with well over 300,000 new jobs created during calendar year 2017 and again in 2018. Almost one of every eight Americans is employed by the healthcare industry.[22] Projections from the US Bureau of Labor Statistics for the next ten years indicate that most of the jobs with the highest anticipated growth are related to health, including personal care aides (754,000), home health aides (425,000), physician assistants (40,000), nurse practitioners (56,000), and physical therapy assistants (27,000). Because human resources are by far the most important driver of healthcare costs, now exceeding $3.5 trillion in the United States per year, you can imagine that people are thinking about how AI can automate operations and alleviate this unbridled growth and related costs. As Katherine Baicker of Harvard put it, "The goal of increasing jobs in health care is incompatible with the goal of keeping health care affordable."[23]

Some economists have suggested that the growth of new job types in healthcare will match or exceed the rate at which AI can replace them. But Kai-Fu Lee, a leading authority in AI, thinks otherwise: "It will soon be obvious that half our tasks can be done better at almost no cost by AI. This will be the fastest transition humankind has experienced, and we're not ready for it."[24]

Hospitals, clinics, and health systems employ people to abstract their medical records to come up with the right billing codes for insurers, and they employ significant numbers of dedicated personnel for payment collection and claims management. The American Academy of Professional Coders has more than 175,000 members with an average salary of $50,000 to do medical bill coding. It is remarkable that the cost of doing the billing for a visit with a doctor in the United States is over $20, or 15 percent of the total cost. It's even worse for an emergency room visit, as the cost of billing accounts for more than 25 percent of the revenue.[25] Overall, in the United States, more than 20 percent of healthcare spending is related to administration.[26] Manual, human scheduling for operating rooms or staffing all the inpatient and outpatient units in a hospital leads to remarkable inefficiencies. Much of the work that attends to patients calling in to schedule appointments could be accomplished with natural-language processing, using human interface as a backup. Algorithms are already being used at some health systems to predict no-shows for clinic appointments, a significant source of inefficiency because missed appointments create so many idle personnel. Even the use of voice assistants to replace or complement nurse call buttons in hospitals by Inovia's AIVA may help improve productivity.[27]

All these operational positions await AI engagement and efficiency upgrades. Some efforts are being made already. One example is Qventus, which uses multimodal data—from the EHR, staffing, scheduling, billing systems, and nurse call lights—to predict proceedings in a hospital's emergency department, operating rooms, or pharmacy. The company claims to have achieved a marked reduction in patients falling in a hospital,[28] the percentage of patients who leave the emergency room without being seen, and the time it takes for a doctor to see a patient.[29] Companies such as Conversa Health, Ayasdi, Pieces Tech, and Jvion are also using AI to take on these logistical tasks, along with many unmet needs for improving efficiency and patient engagement.[30]

How AI can ease medical workflow is exemplified by a program that MedStar Health, the largest health system in the Washington, DC, region, has initiated in its emergency rooms. The typical ER patient has about sixty documents in his or her medical history, which takes considerable time for clinicians to review and ingest. MedStar developed a machine learning system that rapidly scans the complete patient record and provides recommendations regarding the patient's presenting symptoms, freeing doctors and nurses to render care for their patients.[31] Another example is AI automation of medical images, which isn't simply about reading MRIs. The FDA-approved Arterys algorithm called Deep Ventricle enables rapid analysis of the heart's blood flow, reducing a task that can take an hour as blood is drawn and measured by hand to a fifteen-second scan.

Marked improvement in the workflow of medical scans is starting to take hold. Using deep learning algorithms for image reconstruction, multiple reports have indicated the time requirement for acquiring and processing the scans has fallen, improving the quality of the images produced, and substantially reducing the dose of ionizing radiation will become possible. When such improvements are implemented, this may well be one of the first ways we'll see AI promoting safety, convenience, and the potential to lower cost.[32] Another is with radiotherapy for cancer. Researchers at the University College London and DeepMind used an automated deep learning algorithm to markedly accelerate the segmentation processing of scans, achieving similar performance to experienced radiation oncologists for patients with head and neck cancer with remarkable time savings.[33] The use of deep learning algorithms for image segmentation has considerable promise to improve both accuracy and workflow for scans, compared with our prior reliance on traditional algorithms and human expert oversight.

Better prediction of an important diagnosis in real time is another direction of AI efforts, as we've seen, and this issue is of huge importance in hospitals, as one of the major challenges that hospitals face is treating infections that patients catch while hospitalized. Sepsis, a deadly infection common in hospitals, is responsible for 10 percent of intensive care unit admissions in the United States. Treating it costs more than $10 billion per year, and treatment often fails: sepsis accounts for 20 to 30 percent of all deaths among hospitalized patients in the United States. Timely diagnosis is essential since patients can deteriorate very quickly, often before appropriate antibiotics can be selected, let alone be administered and take effect. One retrospective study by Suchi Saria at Johns Hopkins Medicine used data from 53,000 hospitalized patients with documented sepsis, along with their vital signs, electronic medical records, labs, and demographics, to see whether the condition could be detected sooner than it had been. Unfortunately, the accuracy of the algorithm (ROC ~.70) was not particularly encouraging.[34] A second deadly hospital-acquired infection, *Clostridium difficile* or *C. diff*, is also a target of AI. The data to date look a bit more positive. *C. diff* kills about 30,000 people each year in the United States, out of more than 450,000 patients diagnosed.[35] Erica Shenoy and Jenna Wiens developed an algorithm to predict the risk from 374,000 hospitalized patients at two large hospitals using more than 4,000 structured EHR variables for each. Their ROCs were 0.82 and 0.75 for the two hospitals, with many features that were specific to each institution.[36] With automated alerts to clinicians of high *C. diff* risk, it is hoped that the incidence of this life-threatening infection can be reduced in the future.

Preventing nosocomial infections, which one in every twenty-five patients will acquire from a caregiver or the environment, is also an important challenge for hospitals. For example, we know that lack of or suboptimal handwashing is a significant determinant of hospital-acquired infections. In a paper titled "Towards Vision-Based Smart Hospitals," Albert Haque and colleagues at Stanford University used deep learning and machine vision to unobtrusively track the hand hygiene of clinicians and surgeons at Stanford University hospital with video footage and depth sensors. The technology was able to quantify how clean their hands were with accuracy levels exceeding 95 percent (Figure 9.1).[37] Such sensors, which use infrared light to develop silhouette images based on the distance between the sensors and their targets, could be installed in hospital hallways, operating rooms, and at patient bedsides in the future to exploit computer vision's vigilance.

Indeed, machine vision has particular promise for deep learning patterns in the dynamic, visual world of hospitals. The intensive care unit is another prime target for machine vision support. Reinforcement learning has been used as a data-driven means to automate weaning of patients from mechanical ventilation, which previously has been a laborious and erratic clinically managed process.[38]

Surveillance videos of patients could help determine whether there is risk of a patient pulling out their endotracheal (breathing) tube and other parameters not captured by vital signs, reducing the burden on the nurse for detection. The ICU Intervene DNN, from MIT's CSAIL, helps doctors predict when a patient will need mechanical ventilation or vasopressors and fluid boluses to support blood pressure, along with other interventions.[39] Another CSAIL algorithm helps determine optimal time of transfer out of the ICU, with the objective of reducing hospital stay and preventing mortality.[40] Other efforts centered on the ICU reduce the burden on the nurse by automated surveillance with cameras or algorithmic processing of vital signs.

We're still in the early days of machine vision with ambient sensors, but there is promise that this form of AI can be useful to improve patient safety and efficiency. Another common hospital task that ma-

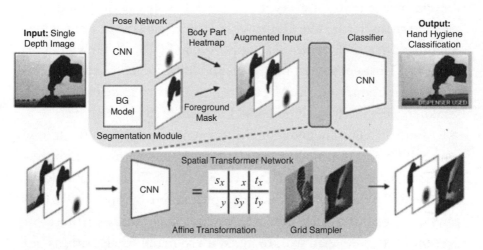

FIGURE 9.1: Handwashing activity classified by machine vision. Source: Adapted from A. Haque et al., *Towards Vision-Based Smart Hospitals: A System for Tracking and Monitoring Hand Hygiene Compliance,* arXiv (2017): https://arxiv.org/abs/1708.00163.

chine vision will likely have a role in changing is placing a central venous catheter, commonly known as a central line, into a patient. Because these lines are so invasive, they carry a significant risk of infection and complications such as a collapsed lung or injury to a major artery. By monitoring proper technique, with respect to both sterile conditions and line placement, safety may improve. Operating rooms could change as machine vision systems continuously track personnel and instruments along with workflow.[41] Prevention of falls in the hospital by cueing into risky patient movements or unsteadiness is also being pursued with AI vision.

A similar story for automated alerts to speed diagnosis and treatment is now ongoing for stroke. The FDA has approved algorithms, developed by Viz.ai, that analyze CT brain images for signs of stroke, enabling neurologists and healthcare teams to rapidly learn whether and what type of stroke has occurred in a patient undergoing scanning. Treatments for reducing the toll of brain damage, including dissolution or removal of clots (thrombectomy), have been validated, so this AI tool is helping to hasten the time to treatment for certain strokes suitable for intervention. That's a critical goal: we lose about

2 million brain cells for every minute a clot obstructs the blood supply.[42] Even earlier in the diagnosis of stroke, paramedics can apply the Lucid Robotic System, FDA approved in 2018, which is a device put on the patient's head that transmits ultrasound waves (via the ear) to the brain, and by AI pattern recognition it helps diagnose stroke to alert the receiving hospital for potential clot removal.[43]

Another major change that will come to the medical workflow, both within and outside hospitals, is how AI will empower nonphysicians to take on more work. There are about 700,000 practicing physicians in the United States complemented by about 100,000 physician assistants and 240,000 nurse practitioners—almost 50 percent of the physician workforce. With so many AI algorithms being developed to support clinicians, it is natural to assume that there will be a more level playing field in the future for these three different groups and that PAs and NPs will be taking on a larger role in the years ahead.[44] The critical assessment of deployment of AI in health systems deserves mention; it will require user research, well-designed systems, and thoughtful delivery of decisions based on models that include risk and benefit. This is unlike the rollout of EHRs into clinical medicine, when many of these vital steps were not incorporated and had serious adverse impact on the day-to-day care of patients.

MAKING HOSPITAL ROOMS OBSOLETE

We get even bolder with the planned "extinction" of the hospital, at least as we know it today.[45] Although we clearly need ICUs, operating rooms, and emergency rooms, the regular hospital room, which makes up the bulk of hospitals today, is highly vulnerable to replacement. Mercy Hospital's Virtual Care Center in St. Louis gives a glimpse of the future.[46] There are nurses and doctors; they're talking to patients, looking at monitors with graphs of all the data from each patient and responding to alarms. But there are no beds. This is the first virtual hospital in the United States, opened in 2015 at a cost of $300 million to build. The patients may be in intensive care units or in their own bedroom, under simple, careful observation or intense scrutiny, but they're all monitored remotely. Even if a patient isn't having any symptoms, the AI surveillance algorithms can pick up a warning and alert the clinician. Their use of high-tech algorithms to remotely detect possible sepsis or heart decompensation, in real time, before such conditions are diagnosed, is alluring. Although being observed from a distance may sound cold, in practice it hasn't been; a concept of engendering "touchless warmth" has taken hold. Nurses at the Virtual Care Center have regular, individualized interactions with many patients over extended periods, and patients say about the nurses that they feel like they "have fifty grandparents now."[47]

Apart from elderly patients with an acute illness, there is a concentrated effort to use AI to support seniors' ability to live and thrive in their home, rather than having to move into assisted living facilities or even needing to have caregivers make frequent visits. There's an extraordinary array of start-ups developing sensors and algorithms that monitor gait, pulse, temperature, mood, cognition, physical activity, and more. Moreover, AI tools to improve vision and hearing can even augment seniors' sensory perception, which would promote their safety and improve their quality of life. For example, with the Aipoly app, a senior with significant visual impairment can simply point to an object with a smartphone and AI will quickly kick in with a voice response identification. It does the same for describing colors. Sensors that can detect whether someone has fallen can be embedded in the floor. And robot assistants in the form of pets as well as specially designed Alexa-like voice assistants like ElliQ (by Startup Robotics) are examples of hardware AI to promote independent living.[48]

Remote monitoring has the potential for very broad use in the future. With each night in the hospital accruing an average charge of $4,700, the economic rationale for providing patients equipment and data plans is not hard to justify. Add to that the comfort of one's home without the risk of acquiring a nosocomial infection or experiencing a sleepless night with constant alarms beeping. Nevertheless, the St. Louis center is pretty much one of a kind right now, and there is little movement to make this the preferred path for patients not requiring an ICU bed. Several issues are holding us back. Some are technological and regulatory. Although systems to monitor all vital signs automatically, such as the Visi device of Sotera Wireless, are approved and currently being used by many health systems, there is no FDA-approved device for home use yet. Until we have FDA devices approved for home use that are automatic, accurate, inexpensive, and integrate with remote monitoring facilities, we've got an obstacle. Perhaps more important in the short term is the lack of reimbursement models for such monitoring and the protracted delays that are encountered with getting new codes set up and approved by Medicare and private insurers.

INSURERS AND EMPLOYERS

Even if there are reimbursement issues standing in the way of medical AI in the home, the dominant forces in control—the health insurers and employers—have much potentially to gain from incorporating AI. Their motive is straightforward: to reduce costs. The public has a very negative sentiment toward insurers because all too often patients encounter denials for services that they perceive are needed or are granted woefully incomplete insurance coverage for the care that they do receive. We surely don't need new algorithms to deny payment or cut those insured short, even though that is a way that AI may find its place in this domain.

As an advisor to the Blue Cross Blue Shield Association, the national headquarters that oversees all thirty-seven regional plans in the United States, I have seen AI beginning to be embraced for select functions. These include making smarter algorithms for patients with diabetes that are not just rules based (as are the only ones operational today) but are anticipatory and factor in deep learning for the individual, including key covariates like the person's daily weight, sleep, nutrition, stress, and physical activity. Their collaboration with

Onduo, a company dedicated to developing such algorithms to fight diabetes, reflects that interest. Indeed, because diabetes is the most common expensive chronic condition, some plans have engaged with other companies to provide virtual coaching services to achieve optimal control and management of diabetes, like Virta or Livongo (see Chapter 11).

When I visited with the UnitedHealth Group leadership at the end of 2017, David Wichmann, the new CEO, showed me how the organization was embracing AI for many specific applications. The company was using natural-language processing to replace keyboarding during office visits in an active initiative, and I saw a demo that convinced me this will be attainable. A UnitedHealth-branded Amazon Echo for various health-related functions was another example of its intended use of the AI voice platform, which was also the case with Blue Cross. In parallel, United has invested heavily in advanced diabetes management companies, acquiring Savvysherpa at the end of 2017, which has developed algorithms for management of type 2 diabetes using continuous glucose sensors, with attendant better outcomes (as far as glucose regulation) and lower costs.

Blue Cross and UnitedHealth are the two largest health insurers in the United States. With their enormous size, covering almost 170 million Americans between them, they tend to move slowly to adapt new technology. While they are certainly using AI tools to make their own business operations run more efficiently as well as those of their subsidiary health services and Big Data efforts (such as Optum Health), the insurance companies are most interested in how AI is implemented in actual healthcare, not in the back office. Surely, though, back-office functions will build further going forward. Accolade Health is changing the way it handles customer service, for example, with a smartphone app called Health Assistant that provides personalized navigation through the health system, ranging from finding a provider for a consultation to dealing with billing and insurance issues.

There are real concerns that AI will present problems to patients as the technology does become a part of insurance operations.

What is particularly worrisome is the potential use of AI analytics to partition populations of patients according to the health risk of each individual and raise individual rates for coverage. In the era of improved prediction of health, there will need to be regulation to avoid discrimination against individuals based on risk. It took many years to get federal legislation to protect individuals against genetic discrimination by employers and health insurers, and even that remains incomplete, as life insurance and long-term disability plans can discriminate based on genetic information. And, although the Affordable Care Act made provisions for excluding preexisting conditions for coverage considerations, that isn't set in stone, as the Trump administration has made clear.[49] Along these lines, risk prediction, by individual, is the next frontier of concerns that remain to be addressed.

Perhaps less pernicious but still worrisome is reliance on "wellness" programs, which most medium to large employers in the United States have, despite the fact that, overall, they have not been validated to promote health outcomes. Typically, a wellness program combines step counting, weight and blood pressure readings, and cholesterol lab tests, as well as some incentive for employees to participate (such as a surcharge on an employee's contribution to the cost of insurance). But wellness is poorly defined, and the cost effectiveness of such strategies has been seriously questioned.[50] One way such programs could be improved, however, is through the use of virtual medical coaches, which could gather and make use of far more granular and deeper information about each individual. Here again is the concern that employers, through their insurance providers, could use such data to financially disadvantage individuals, which could be a major disincentive for patients to use such technology.

Looking overseas, one relatively small insurer that has been gaining some experience with more comprehensive data is Discovery Limited, which originated in South Africa but is also now available in Australia, China, Singapore, and the UK. Its Vitality program uses a

Big Data approach to capture and analyze physical activity, nutrition, labs, blood pressure, and, more recently, whole genome sequences for some individuals. There have yet to be any publications regarding the betterment of health outcomes with this added layer of data, but it may represent a trend for insurers in the future.

MEDICAL AI AT THE NATIONAL LEVEL

AI in medicine hasn't attracted quite the same level of attention or ambition as AI for global military, cyber, and superpower dominance, about which Vladimir Putin declared, "Whoever becomes the leader in this sphere will become the ruler of the world."[51] The goal is better health and lower cost for citizens, not world dominance. It is still happening around the world. Canada has been an epicenter for deep learning, with Geoffrey Hinton and colleagues at the University of Toronto, and the dozens of former students who now have prominent AI leadership roles at Google, Uber, Facebook, Apple, and other leading tech companies. Hinton believes that AI will revolutionize healthcare, and his company Vector is using neural networks for the massive datasets throughout hospitals in Toronto. His Peter Munk Cardiac Centre is focused on cardiovascular care, using AI to actualize remote monitoring of patients. Deep Genomics, started by Brendan Frey, one of Hinton's students, is using AI for genomic interpretations.[52] These are just a few of the AI healthcare initiatives and companies in Canada.

It's likely the big changes in AI medicine will take hold far more readily outside the United States, and countries like India and China are particularly likely to be prominent first movers. India has a doctor-to-patient ratio of only 0.7 per 1,000 people, which is less than half that of China (1.5) and substantially less than that of the United States (at 2.5). The ingenuity in AI in India is reflected by companies like Tricog Health for cloud-based heart condition diagnosis, Aindra Systems for automated detection of cervical cancer via path samples, Niramai for early breast cancer detection, and Ten3T for

remote monitoring. The pioneering work at the Aravind Eye Hospitals, the largest eye care network in the world, in collaboration with
Google, was the basis for deep learning algorithms to detect diabetic
retinopathy—a condition for which more than 400 million people
are at risk but the majority of whom are not getting screened.[53]

But it's China that seems positioned to take the lead in AI for
medicine. So many important factors line up: unparalleled quantity
of data collection (citizens cannot opt out of data collection), major government and venture fund investment, major AI programs
at most of the big universities, and a very supportive regulatory environment.[54] Beyond all these attributes, there is the need. As Lin
Chenxi of Yitu, a Chinese medical image recognition company, put
it: "In China, medical resources are very scarce and unequally distrib-
uted so that the top resources are concentrated in provincial capitals.
With this system, if it can be used at hospitals in rural cities, then
it will make the medical experience much better."[55] There are only
twenty eye doctors for every million people in China, recapitulating
the general trend for the country, where the proportions of various
specialists to the population as a whole are one-third or less the proportions found in the United States. China is using more than 130
medical AI companies to promote efficiency and expand access in its
healthcare system.[56]

Behind all this is major support. In 2018, the Chinese government issued its manifesto to become the world leader in AI, making
the endeavor its own version of an Apollo 11 moon shot.[57] Although
there has been a talent gap between China and the United States,
with the United States having a great many more computer scientists with AI expertise, that gap is quickly closing. Since 2014, more
research papers have been published on DNN in China than in the
United States. China is now second to the United States in patent
applications and in private investment for AI.[58] The parallels in the
tech titan oligopoly between China and the United States are striking. Tencent is like Facebook, Baidu like Google, and Alibaba similar
to Amazon. While Chinese AI achievements may not garner as much

international publicity as those that happen in the United States, China's advancements in image and speech recognition have been noteworthy.

The results so far in medicine are striking. The Guangzhou Hospital is using AI, trained from 300 million records (no wonder the *Economist* characterized China as "the Saudi Arabia of data") from patients across the country, for almost every part of its operation—organizing patient records, suggesting diagnoses via a WeChat bot interaction, identifying patients through facial recognition, interpreting CT scans, and operating room workflow.[59] Tencent is very active in medical image diagnosis and drug discovery, and has backed the WeDoctor Group, a hospital of the future initiative. VoxelCloud, an eye-imaging interpretation company also supported by Tencent, is deploying diabetic retinopathy AI screening broadly to counter the leading cause of blindness among China's working age population. The AI company that has gone most intensively into medicine to date is iFlytek, which is a major global player in speech recognition. In 2018, it launched an AI-powered robot called Xiaoyi that has passed China's medical licensing examination for human physicians (with a score of 456, which was 96 points past the required level).[60] With iFlytek's robot's ability to ingest and analyze individual patient data, it plans to integrate these capabilities with general practitioners and cancer doctors throughout China. The start-up PereDoc, founded by cancer radiologist Chongchong Wu, has already installed its medical imaging AI algorithms in twenty Chinese hospitals.[61] The company Ant Financial has a chatbot that has exceeded human performance for customer satisfaction.[62] Ant Financial also has acquired the US company EyeVerify (now renamed Zoloz), which makes eye recognition AI algorithms. With this sampling of healthcare AI getting its legs in China, concern has also been expressed about the surveillance state and the potential breach of data privacy. For example, Ant Financial's three-digit credit score might be linked to medical data. Along with omnipresent video cameras installed every hundred meters in

most cities, each citizen's ID number may be linked with facial recognition, DNA information, iris scan, and other biometric data.[63]
These extensive measures of AI recognition and surveillance have yet
to be coupled with any betterment of health outcomes.

India and China are not alone in pushing forward medical AI.
France and the United Kingdom both have committed significant
resources to advance AI, and both have specified priorities and objectives for AI in healthcare. Just after issuing a French government
policy statement "For a Meaningful Artificial Intelligence" and
significant (near $2 billion) investment in 2018, Emmanuel Ma-
cron gave an interview to *Wired* magazine.[64] The editor, Nicholas
Thompson, asked him, "What was the example of how AI works
that struck you the most and that made you think, 'Ok, this is going
to be really, really important'?" Macron said, "Probably in healthcare—where you have this personalized and preventive medicine
and treatment. We had some innovations that I saw several times in
medicine to predict, via better analysis, the diseases you may have
in the future and prevent them or better treat you. . . . Innovation
that artificial intelligence brings into healthcare systems can totally
change things: with new ways to treat people, to prevent various
diseases, and a way—not to replace the doctors—but to reduce the
potential risk."[65]

The United Kingdom is also betting big on AI's future and emphasizing healthcare. When the UK government issued four Grand
Challenges, one centered on medicine, Theresa May declared, "The
development of smart technologies to analyse great quantities of data
quickly and with a higher degree of accuracy than is possible by human beings, opens up a whole new field of medical research and
gives us a new weapon in our armory in the fight against disease."[66]
In 2018, I was commissioned by the UK to work with the National
Health Service on planning the future of its healthcare, particularly
leading a review on the impact of AI and other medical technologies
on its workforce over the next two decades.[67] The opportunity to

work with leaders of AI, digital medicine, genomics, and robotics, along with ethicists, economists, and educators was an extraordinary experience in the context of a single-payer healthcare system with a palpable will to change and adapt. The full report was issued in 2019, where we project major impacts at every level—the patient, the clinicians, and the health systems throughout the country.

While I've taken a country-by-country approach here, I dream about coalescing medical data across all countries someday. Going global is the best way to achieve medical AI's greatest potential—a planetary health knowledge resource, representing the ultimate learning health system. It would compensate for the fact that most biomedical research performed to date has been done with subjects of European ancestry, which means that physicians often cannot extrapolate their findings to individuals of other ancestries. If all members of the species had comprehensive data in such a resource, with their treatments and outcomes, this would enable AI nearest neighbor analysis to find "digital twins." These are individuals who most resemble, by all demographic, biologic, physiologic, and anatomic criteria, the person at risk or with a new important diagnosis. Knowledge of outcomes from twins would enable better prevention or treatment of the individual and the next generation. The likelihood of assembling such a resource for the world's population is very low, especially impaired by concerns over privacy, data security, and cross-cultural sharing considerations. But we are seeing this at a smaller scale, with efforts such as Tempus Labs in cancer (Chapter 7). It's a think-big scenario to imagine what awaits us in the longer term for all medical conditions without geographic boundaries. But even if the odds are low now, I hope recognition of the possibilities will help make those odds better. As soon as patient outcomes are shown to be unequivocally improved by having digital twins inform best treatment, it is likely there will be substantial commitments across health systems to develop and prioritize such infrastructure.

With this review of the opportunities at the level of healthcare systems, it's time to turn upstream—to the discovery side of drugs and the science that leads to better treatments and mechanistic insights about health and disease. AI is beginning to have a big impact there, too, which over time may further improve the outcomes and efficiency of medical practice.

chapter ten

DEEP DISCOVERY

People think technology + big data + machine learning = science. And it's not.

—John Krakaeur

THE MASSIVE DATASETS THAT ARE NOW EMERGING IN BIO-medicine have created an imperative to adopt machine learning and AI. Take, for example, the Cancer Genome Atlas of multi-dimensional biologic data, comprising various "omics" (genomics, proteomics, and so on). All told the atlas holds more than 2.5 peta-bytes of data generated from more than 30,000 patients.[1] No human could wade through that much data. As Robert Darnell, an oncologist and neuroscientist at Rockefeller University put it, "We can only do so much as biologists to show what underlies diseases like autism. The power of machines to ask a trillion questions where a scientist can ask just ten is a game-changer."

That said, unlike the immediate and ongoing changes that AI is unleashing on clinicians in the pattern-heavy medical fields like pathology and radiology, AI isn't yet challenging the status quo for scientists in any significant way; AI is just here to help. As Tim Appenzeller put it in the journal *Science,* AI is still "the scientists' apprentice." But the help that AI can give is very powerful; in 2017, the cover of *Science* trumpeted, "AI Transforms Science." Not only has it "spawned AI neuroscience" (as we'll soon see), it has "supercharged the process of discovery." Indeed, *Science* saw something more on the horizon: "the prospect of fully automated science," and the journal claimed that "the tireless apprentice may soon become a full-fledged colleague."[2]

AI colleagues still seem like a distant prospect to me, but regardless of whether AI ever displaces scientists, AI science and discovery efforts are moving fast. Indeed, AI has been developing for life science applications at a much faster clip than it has for healthcare delivery. After all, basic science does not necessarily require validation from clinical trials. Nor does it need acceptance and implementation by the medical community, or oversight by regulatory agencies. Even though all the science has not yet made it to the clinic, ultimately, these advances will have major impact in how medicine is practiced, be it by more efficient drug discovery or elucidation of biologic pathways that account for health and disease. Let's see what the apprentice has been up to.

THE BIOLOGIC OMICS AND CANCER

For genomics and biology, AI is increasingly providing a partnership for scientists that exploits the eyes of machines, seeing things that researchers couldn't visualize, and sifting through rich datasets in ways that are not humanly possible.

The data-rich field of genomics is well suited for machine help. Every one of us is a treasure trove of genetic data, as we all have 6 billion letters—A, C, G, and T—in our diploid (maternal and paternal copies) genome, 98.5 percent of which doesn't code for proteins. Well more than a decade after we had the first solid map of a human genome, the function of all that material remains elusive.

One of the early deep learning genomic initiatives, called Deep-SEA, was dedicated to identifying the noncoding element's function.

In 2015, Jian Zhou and Olga Troyanskaya at Princeton University published an algorithm that, having been trained from the findings of major projects that had cataloged tens of thousands of noncoding DNA letters, was able to predict how a sequence of DNA interacts with chromatin. Chromatin is made up of large molecules that help pack DNA for storage and unravel it for transcription into RNA and ultimately translation into proteins, so interactions between chromatin and DNA sequences give those sequences an important regulatory role. Xiaohui Xie, a computer scientist at UC Irvine, called it "a milestone in applying deep learning to genomics."[3]

Another early proof of this concept was an investigation of the genomics of autism spectrum disorder. Before the work was undertaken, only sixty-five genes had been linked to autism with strong evidence. The algorithms identified 2,500 genes that were likely contributory to or even causative of the symptoms of the autism spectrum. The algorithm was also able to map the gene interactions responsible.[4]

Deep learning is also helping with the fundamental task of interpreting the variants identified in a human genome after it has been sequenced. The most widely used tool has been the genome analysis tool kit, known as GATK. In late 2017 Google Brain introduced DeepVariant to complement GATK and other previously existing tools. Instead of using statistical approaches to spot mutations and errors and figure out which letters are yours or artifacts, DeepVariant creates visualizations known as "pileup images" of baseline reference genomes to train a convolutional neural network, and then it creates visualizations of newly sequenced genomes in which the scientists wish to identify variants. The approach outperformed GATK for accuracy and consistency of the sequence. Unfortunately, although DeepVariant is open source, it's not eminently scalable at this point because of the expense of its heavy computational burden that requires double the CPU core-hours compared with GATK.[5]

Determining whether a variant is potentially pathogenic is a challenge, and when it's in a noncoding region of the genome it gets even more difficult. Even though there are now more than ten AI algorithms to help with this arduous task, identifying disease-causing variants remains one of the most important unmet needs. The same Princeton team mentioned previously took deep learning of genomics another step forward by predicting noncoding element variant effects on gene expression and disease risk.[6] A team led by the Illumina genomics company used deep learning of nonhuman primate genomes to improve the accuracy of predicting human disease-causing mutations.[7]

Genomics (DNA) is not the only omic field that is ripe for deep and machine learning. Deep learning has already been applied to every layer of biological information, including gene expression; transcription factors and RNA-binding proteins; proteomics; metagenomics, especially the gut microbiome; and single-cell data.[8] DeepSequence and DeepVariant are AI tools to understand the functional effect of mutations or make accurate calls of genomic variants, respectively, both of which surpass performance of previous models.[9] DeepBind is used to predict transcription factors. DeFine quantifies transcription factor–DNA binding and helps assess the functionality of noncoding variants. Other efforts have predicted specificities of DNA- and RNA-binding proteins, protein backbones from protein sequences, and DNase I hypersensitivity across multiple cell types.[10] The epigenome has been analyzed by DeepCpG for single cell methylation states,[11] chromatin marks and methylation states have been predicted,[12] and deep learning neural networks have improved upon the challenging analysis of single cell RNA-sequence data.[13] Within and between each omic layer, the interactions are seemingly infinite, and machine learning is increasingly being applied to help understand the myriad ways of how genes interact even within a single cell.[14]

The combination of AI with genome editing has proven to be especially formidable. Microsoft Research developed an algorithmic approach called Elevation, shown to predict off-target effects across the human genome when attempting to edit DNA and so to predict the optimal place to edit a strand of DNA and to design guide RNAs for CRISPR editing (the acronym stands for snippets of DNA, or more formally "clustered regularly interspaced short palindromic repeats").[15] It outperformed several other CRISPR design algorithms, many of which use machine learning. While critical to precision in experimental biology, such algorithms will play a key role in the many clinical trials moving forward that use CRISPR system editing for diseases like hemophilia, sickle cell, and thalassemia.

Perhaps unsurprisingly, given that it is one of machine learning's core strengths, image recognition is playing a critical role in cell analysis: to sort shape, classify type, determine lineage, identify rare cells in the blood, or distinguish whether cells are alive or dead.[16] The inner workings of cells have been the focus of DCell, a deep learning algorithm that predicts growth, gene-gene interactions, and other functions.[17]

Cancer is a genomic disease, so it's no wonder that oncology has particularly benefited from this injection of AI. Beyond help in the interpretation of sequencing data from tumors, such as has been done for glioblastoma, a brain cancer, we've seen new insights to the genesis and biophysics of cancer.[18]

A tumor's DNA methylation data has also proven to be very useful input for AI classification in cancer. Pathologists characteristically use histological samples on slides to diagnose brain tumors. There are challenges to doing this: there are many rare cancers, which creates enough of a challenge if a pathologist hasn't seen it before; the cells in a tumor are a mosaic of different types; a biopsy is usually an incomplete sample of the cell types in a tumor; and visually reviewing a slide is inevitably subjective. In a seminal 2018 study by David Capper at the Charité hospital in Berlin and colleagues, whole genome methylation of tumor specimens was shown to be 93 percent accurate in classifying all eighty-two classes of brain cancer, which far exceeded the accuracy of pathologists. The machine-determined DNA methylation status then led to recategorization of more than 70 percent of human-labeled tumors, which could mean significantly different predictions about prognosis and decisions about treatment.[19] Such findings have important implications for both cancer biology lab experiments and medical practice.

We've learned quite a bit about the evolution of cancer with the help of AI. Hidden signals of the evolution trajectory of cancer in 178 patients became discernible via a transfer learning algorithm, with implications for patient prognosis.[20] But in the world of AI hype this was translated to the front page of the *Express,* in the United Kingdom, where the research was conducted, as "Robot War on Cancer."[21] AI tools have helped in the discovery of cancer somatic mutations[22] and to understand the complexity of cancer gene interactions.[23]

A last notable example of exploring cancer with AI used a complex biological system to predict whether cells would become cancerous. Using a frog-tadpole model of tumor development, researchers treated populations of tadpoles with a combination of three reagents to find combinations that caused the melanocytes of some of the frog larvae to develop into a cancer-like form. Despite the fact that at a population level not all tadpoles developed cancer, the researchers were surprised by the fact that all the melanocytes in a given tadpole behaved the same way, either all becoming cancerous or all developing normally. The researchers sought to identify a combination of reagents that would cause an intermediate form, in which only some cells within an organism became cancerous, to develop.

Having performed a few studies to develop the ground truth, they then used AI models to run 576 virtual experiments, computationally simulating the embryo development under a range of combinations of reagents. All but one failed. But in the AI haystack was a model that predicted a cancer-like phenotype, where not all cells develop the same way, and this was prospectively fully validated. The study's author, Daniel Lobo of University of Maryland, Baltimore County, said, "Even with the full model describing the exact mechanism that controls the system, a human scientist alone would not have been able to find the exact combination of drugs that would result in the desired outcome. This provides proof-of-concept of how an artificial intelligence system can help us find the exact interventions necessary to obtain a specific result."[24]

DRUG DISCOVERY AND DEVELOPMENT

Successfully identifying and validating a novel drug candidate is one of the greatest challenges in biomedicine, and it is certainly one of the most expensive. The awesome cost and high risk of failure has encouraged a rapid embrace of any technology that promised to reduce the expense or difficulty of developing drugs. A decade ago there was a heavy investment in hardware—robotics to perform high-throughput and mass screening of molecules; now the emphasis is on algorithmic automation. By 2018 there were more than sixty start-ups and sixteen pharmaceutical companies using AI for drug discovery.[25] Much like the researchers who investigated tadpole cancer, these groups deploy multiple AI tools to find the needle in the haystack, whether by raking through the biomedical literature, mining millions of molecular structures in silico, predicting off-target effects and toxicity, or analyzing cellular assays at massive scale. Faster development of more potent molecules is in the works (automating molecule design). There is even hope and preliminary data that AI chemical screening will markedly reduce the need for preclinical animal testing.[26] The AI strategy for these companies is remarkably heterogeneous so I will briefly review a sampling of them so you can see the potential impact of AI (see Table 10.1).[27]

Use of natural-language processing to ingest all of what is known about drugs and molecules through the biomedical literature and chemical databases is a start. Analyzing all that data in a hypothesis-free mode without human bias (such as pet theories) is another advantage.

We've heard that there are more stars in the sky than grains of sand on Earth. The galactic scale also applies to small molecules. There are

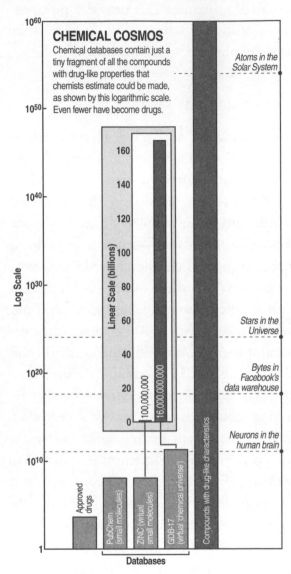

FIGURE 10.1: Comparing chemical databases with other metrics on a logarithmic scale. Source: Adapted from A. Mullard, "The Drug-Maker's Guide to the Galaxy," *Nature* (2017): 549(7673), 445–447.

about 10^{60} chemical compounds with drug-like features that can be made, far more small molecules than the number of atoms in the solar system (Figure 10.1).[28] That's the perfect substrate for AI, and companies like Exscientia are developing full catalogs of these compounds, and Epiodyne collated 100 million compounds that haven't been made but would be easy to synthesize. Such work isn't being done only at start-up companies. UCSF's Brian Shoichet led a painkiller discovery project that narrowed a list of 3 million compounds down to just twenty-three. The organic chemists at University of Münster in Germany have been using deep learning to make the synthesis of compounds far more predictive, quick, and easy.[29] A robot named Eve at the University of Cambridge, endowed with AI library screening capabilities, was able to find multiple lines of evidence for action of an antimalarial drug.[30] Jean-Louis Reymond, at the University of Bern in Switzerland, has put together a database known as GDB-17 of 166 billion compounds, representing all chemically feasible molecules made up of seventeen or fewer atoms. Nearest-neighbor algorithmic analysis can sift through the whole database in just a few minutes to find new molecules that have effects similar to those of known drugs. Many compounds in Reymond's database have turned out to be very hard to synthesize, so he whittled it down to a "short list" of 10 million easy-to-make compounds. Just 10 million!

Predicting chemical reaction via machine learning has advanced, exemplified by a study Abigail Doyle of Princeton University and colleagues published in 2018. She made it seem so simple: "You draw out the structures—the starting materials, catalysts, bases—and the software will figure out shared descriptors (common chemistry features) between all of them. That's your input. The outcome is the yields of the reactions. The machine learning matches all those descriptors to the yields, with the goal that you can put in any structure and it will tell you the outcome of the reaction."[31]

Insilico Medicine is working on cancer drug discovery, screening more than 72 million compounds from a public database, creatively using a pair of generative adversarial neural networks. The first goal is to identify potentially therapeutic molecules and the second deletes those that are based on prior patented compounds.[32]

BenevolentAI, one of the largest private AI companies in Europe, has built natural-language processing that sifts through the biomedical literature and chemical databases. One of the most impressive papers in AI drug discovery to date comes from Marwin

Segler, an organic chemist at BenevolentAI.[33] He and his colleagues at the University of Munster designed a deep learning algorithm to learn on its own how reactions proceed, from millions of examples. It was used to create small organic molecules from more than the 12 million known single-step organic chemistry reactions.[34] They even tested chemists at two prestigious institutes to see, on a double-blind basis, whether they could distinguish between AI and human synthetic reaction routes. And they couldn't. Likewise, Leroy Cronin and his team at the University of Glasgow designed an organic synthesis robot that used machine learning to search for new chemical reactions.[35] The robot can do thirty-six reactions a day; a chemist only three or four. Moreover, the robot did reactions where the outcome couldn't be predicted beforehand.[36] Derek Lowe reflected on this progress: "The idea that intellectual tasks can be categorized as automatable grunt work will probably be insulting to many chemists, and will certainly feel like a threat. But the use of artificial intelligence will actually free up time in which they can think more about the higher-level questions of which molecules should be made and why, rather than focusing on the details of how to make molecules."[37]

Recursion Pharmaceuticals, an image-processing company, uses algorithms and automated microscopes to interrogate high-throughput drug testing of human cells, getting into such granular details as cell and nucleus size and shape. More than 2,000 molecules were modeled to see which could convert sick cells, modeling genetic diseases, into healthy-looking cells.[38] The company has identified at least fifteen new potential treatments with this strategy, and one is moving forward in clinical trials for cerebral cavernous malformation.

As its name implies, Deep Genomics takes a genomic anchoring approach to deep learning. Back in 2014, the Toronto-based group, led by Brendan Frey, published an impressive paper on the human splicing code, yielding thousands of potential targets for patients with diseases such as autism spectrum disorder and spinal muscular atrophy.[39]

AI COMPANY	TECHNOLOGY	PARTNER	INDICATION(S)
Atomwise	DL from molecular structure	Merck	Malaria
BenevolentAI	DL and NLP of research literature	Janssen	Multiple
BERG	DL of biomarkers from patient data	None	Multiple
Exscientia	Biospecific compounds via Bayesian models of ligand activity	Sanofi	Metabolic diseases
GNS Healthcare	Bayesian probabilistic inference for efficacy	Genentech	Oncology
Insilico Medicine	DL from drug and disease databases	None	Age-related diseases
Numerate	DL from phenotypic data	Takeda	Oncology, CNS, and gastroenterology
Recursion	Cellular phenotyping via machine vision analysis	Sanofi	Rare genetic diseases
twoXAR	DL screening from literature and assay data	Santen	Glaucoma

TABLE 10.1: List of selected companies working on AI for drug discovery, DL—deep learning, CNS—central nervous system, NLP—natural-language processing. Source: Adapted from E. Smalley, "AI-Powered Drug Discovery Captures Pharma Interest," *Nat Biotechnol* (2017): 35(7), 604–605.

Atomwise uses deep learning algorithms to screen millions of molecules, which by late 2017 led to more than twenty-seven drug discovery projects treating conditions ranging from Ebola to multiple sclerosis.[40] The company's neural network, along with 3-D models, provides a list of seventy-two drugs with the highest probability of favorably interacting with the molecular underpinnings of a particular disease.[41] As Gisbert Schneider at the Swiss Federal Institute of Technology pointed out, "The concept of automated drug discovery could help to considerably reduce the number of compounds to be tested in a medicinal chemistry project and, at the same time, establish a rational unbiased foundation of adaptive molecular design."[42]

These new approaches have spurred new private-public partner-ships. One, known as ATOM, for Accelerating Therapeutics for Op-portunities in Medicine, brings together multiple academic centers (Duke and Tulane Universities) and pharmaceutical companies (in-cluding Merck, AbbVie, Monsanto) to "develop, test, and validate a multidisciplinary approach to cancer drug discovery in which mod-ern science, technology and engineering, supercomputing simulations, data science, and artificial intelligence are highly integrated into a single drug-discovery platform that can ultimately be shared with the drug development community at large."[43] The goal of ATOM is to reduce how long it takes to go from identifying a potential drug target to developing a drug candidate that hits the target.[44] That's normally a four-year bottleneck. ATOM seeks to turn it into a one-year lag. Project Survival is a public-private consortium funded by BERG Health that collects biological samples from patients with cancer and, in a seven-year initiative, mines the data integrated with clinical information for each patient to promote biomarker discovery and early detection.[45]

The use of AI in this field extends beyond facilitating drug dis-covery to predicting the right dose for experimental drugs. Since the optimal drug dose may depend on so many variables for each in-dividual, such as age, gender, weight, genetics, proteomics, the gut microbiome, and more, it's an ideal subject for modeling and deep learning algorithms. The challenge of getting the dose right is height-ened by the possibility of drug-drug interactions. Already multiple academic centers have used this approach, including UCLA, Stan-ford, UCSF, Virginia Tech, and the University of Kansas. As Josep Bassaganya-Riera of Virginia Tech put it, "Every person is going to have their own certain set of parameters, and we need to understand what that unique mix of characteristics means, rather than analyzing each individual trait. Machine learning helps us to do that."[46]

There is certainly plenty of hype about AI and drug discovery with headlines like "AI Promises Life-Changing Alzheimer's Drug Breakthrough."[47] Or claims like that of BenevolentAI "to cut drug development timelines by four years and improve efficiency by 60 percent compared to pharma industry averages."[48] Only time will tell whether the transformational potential of all these different shots on goal to rev up drug discovery will be actualized.

NEUROSCIENCE

The intersection of AI and brain science is so rich that I could represent it as a complex wiring diagram. But I won't! Neuroscience has provided significant inspiration to AI researchers since the conceptual origin of artificial intelligence—especially the artificial neural networks that remain so common throughout the field. But, as we'll see, the relationship between neuroscience and artificial intelligence—not to mention computer science more generally—has been and continues to be one where knowledge and breakthroughs in one field change the other. In more ways than one, if computer scientists ever come to mean scientists who are computers rather than scientists who study computers, it will be because of that self-reflexive, even symbiotic, relationship.

The use of AI in neuroscience is really taking off. Perhaps surprisingly, much work has been dedicated not to humans' brains but to flies'. I was particularly struck by the work of Alice Robie at the Howard Hughes Medical Institute.[49] Beginning with videos of 400,000 flies, she used machine learning and machine vision methodology to map the triad of gene expression, traits, and their precise anatomical basis. Ultimately, she generated whole-brain maps that correlated both movements, such as walking backward, and social behaviors, such as female aggression, with more than 2,000 genetically targeted populations of brain cells.

Understanding brains also helps us understand problems in computer science. Remarkably, it was the fruit fly again that proved to be instrumental in understanding the fundamental computing problem of "similarity search"—identifying similar images or documents in a large-scale retrieval system.[50] In this case, it wasn't an image or document retrieval but an odor. It turns out that the fly's olfactory system uses three nontraditional computational strategies, whereby learning from the tagging of one odor facilitates recognition of a similar odor. Who would have guessed that nearest-neighbor computing searches would have common threads with the fly's smell algorithm?

A striking achievement in AI for understanding the brain has been to model spatial navigation, the complex cognitive perceptual mapping task—integrating information about our body's movement speed and direction, our place in space. To do this, the brain relies principally on three kinds of neurons: First, there are the place cells, which fire when we are at a particular position. Second, there are the head-direction cells, which signal the head's orientation. Third, and perhaps most remarkable, there are the grid cells, which are arranged in a perfectly hexagonal shape in the hippocampus. The hippocampus is often referred to as the brain's GPS, and the grid cells make clear why. They fire when we are at a set of points forming a hexagonal grid pattern, like a map inside our head that our brains impose upon our perception of the environment.[51]

But until DeepMind researchers delved into this, it was unclear how grid cells really worked. One important question was whether grid cells could help calculate the distance and direction between two points, enabling our brains to select the shortest route from one to the other. This is known as vector-based navigation, and it was a theory without empirical support. To figure out whether vector-based navigation was indeed what our brain did, DeepMind and collaborating computer scientists trained a recurrent neural network to localize simulated rodents in a virtual environment. This led to the spontaneous emergence of hexagonal, grid-like representations, akin to neural activity patterns in mammals, confirming path navigation. Then, using complex virtual reality gaming environments and a deep reinforcement neural network, the artificial agent exhibited superhuman performance, exceeding professional game players, taking shortcuts and novel routes, and demonstrating vector-based navigation. When the grid cells in the network were silenced, the agent's ability to navigate was adversely affected.

This grid cell study gives us a taste of the excitement and advances in neuroscience that are being influenced and unraveled by AI. Christof Koch, who heads up the Allen Institute, provided the endeavor world-historical contextualization: "While the 20th century was the century of physics—think the atomic bomb, the laser and the transistor—this will be the century of the brain. In particular, it will be the century of the human brain—the most complex piece of highly excitable matter in the known universe."[52] We're also seeing how advances in computer science can help us better understand our brains, not just by sorting out the mechanics by which the brain works, but by giving us the conceptual tools to understand how it works. In Chapter 4 I reviewed backpropagation, the way neural networks learn by comparing their output with the desired output and adjusting in reverse order of execution. That critical concept wasn't thought to be biologically plausible. Recent work has actually borne out the brain's way of using backpropagation to implement algorithms.[53] Similarly, most neuroscientists thought biological neural networks, as compared with artificial neural networks, only do supervised learning. But that turns out not to be the case with ample support of reinforcement learning in the prefrontal cortex of the brain. The line between biological and artificial neural networks was blurred with the scaling up of DNA-based neural networks that recognize molecular patterns, expanding markedly from only recognizing the four distinct DNA molecules to classifying nine categories, and creating the potential for embedded learning within autonomous molecular systems.[54]

Using AI to reconstruct neural circuits from electron microscopy represents another example of the interplay. For "connectomics"— the field of comprehensively mapping biological neuronal networks in our nervous system—Google and Max Planck Institute researchers automated the process, improving the accuracy by an order of magnitude.[55]

PROPERTIES	COMPUTER	HUMAN BRAIN
Number of basic units	Up to 10 billion transistors	100 billion neurons 100 trillion synapses
Speed of basic operation	10 billion/sec.	<1,000/sec.
Precision	1 in 4.2 billion (for a 32-bit processor)	1 in 100
Power consumption	100 watts	10 watts
Information processing mode	Mostly serial	Serial and massively parallel
Input/output for each unit	1–3	1,000

TABLE 10.2: The differences in properties between computers and the human brain. Source: Adapted from L. Luo, "Why Is the Human Brain So Efficient?" *Nautil.us* (2018): http://nautil.us/issue/59/connections/why-is-the-human-brain-so-efficient.

Not only is AI playing a major role in how we do neuroscientific research, but neuroscience has long played an important role in the development of AI; as we make further progress in how the brain works, that influence stands to only grow. Both the Perceptron, invented by Frank Rosenblatt, and its heir, the artificial neural network, developed by David Rumelhart, Geoffrey Hinton, and colleagues, were inspired by how biological neurons and networks of them, like the human brain, work. The architecture and functionality of many recent deep learning systems have been inspired by neuroscience.

There are certainly some architectural parallels of neurons and synapses (Figure 10.2) and separate circuits for input, output, central processing, and memory. But the differences are pretty striking, too (Table 10.2). Our energy-efficient brain uses only about 10 watts of power, less than a household light bulb, in a tiny space less than 2 liters, or smaller than a shoebox. The K supercomputer in Japan, by contrast, requires about 10 megawatts of power and occupies more than 1.3 million liters.[56] Where our brain's estimated

100 billion neurons and 100 trillion connections give it a high tolerance for failure—not to mention its astonishing ability to learn both with and without a teacher, from very few examples—even the most powerful computers have poor fault tolerance for any lost circuitry, and they certainly require plenty of programming before they can begin to learn, and then only from millions of examples. Another major difference is that our brain is relatively slow, with computation speeds 10 million times slower than machines, so a machine can respond to a stimulus much faster than we can. For example, when we see something, it takes about 200 milliseconds from the time light hits the retina to go through brain processing and get to conscious perception.[57]

Another important difference between computers and humans is that machines don't generally know how to update their memories and overwrite information that isn't useful. The approach our brains take is called Hebbian learning, following Donald Hebb's maxim that "cells that fire together wire together."[58] The principle explains the fact that if we use knowledge frequently, it doesn't get erased. It works thanks to the phenomenon of synaptic neuroplasticity: a brain's circuit of repetitive, synchronized firing makes that behavior stronger and harder to overwrite.

Until recently, computers didn't ever work this way. Now artificial neural networks have been designed to mimic the functioning of such "memory aware synapses."[59] This was achieved by sequences of object-recognition tasks, like training a network to recognize dogs running and then a person playing sports. In an unsupervised manner, the importance measure of each parameter of the network is accumulated. Then its performance can be established with retesting for seeing dogs running. In such a way, AI is learning what to remember and what it can forget.

This isn't the only way in which our expanding knowledge of the brain is reshaping our understanding of AI and computers. Much as Hinton was inspired by our nervous systems as he worked out the

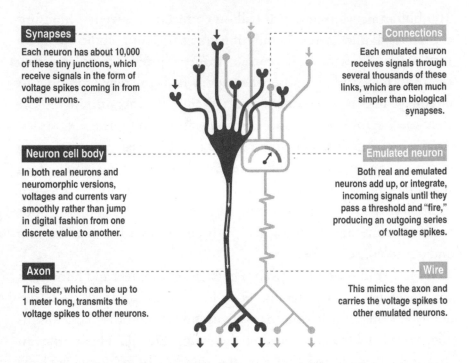

Synapses

Each neuron has about 10,000 of these tiny junctions, which receive signals in the form of voltage spikes coming in from other neurons.

Connections

Each emulated neuron receives signals through several thousands of these links, which are often much simpler than biological synapses.

Neuron cell body

In both real neurons and neuromorphic versions, voltages and currents vary smoothly rather than jump in digital fashion from one discrete value to another.

Emulated neuron

Both real and emulated neurons add up, or integrate, incoming signals until they pass a threshold and "fire," producing an outgoing series of voltage spikes.

Axon

This fiber, which can be up to 1 meter long, transmits the voltage spikes to other neurons.

Wire

This mimics the axon and carries the voltage spikes to other emulated neurons.

FIGURE 10.2: The homology between biological neurons and artificial neurons. Source: Adapted from M. Waldrop, "Neuroelectronics: Smart Connections," *Nature News* (2013): 503(7474), 22–24.

first artificial neural network algorithms, researchers are using our knowledge of the brain to rebuild computers themselves.

For a long time, almost all we knew about the human brain was derived from the study of dead tissue devoid of any electrical activity. The Allen Institute for Brain Science published data from about three hundred live human brain cells derived from surgical samples that, with permission of thirty-six patients, were put on life support to study their structure and function. The resulting three-dimensional maps and better understanding of how neurons interpret incoming signals and generate outgoing ones revealed striking similarities with the functioning of computer neurons.[60]

While this newfound ability to zoom in and reconstruct single, live human brain cells is exciting, not everyone is ready to pronounce this as a big jump forward. David Marr, a British neuroscientist, famously said, "Trying to understand perception by understanding neurons is like trying to understand a bird's flight by studying only its feathers. It just cannot be done." Our real knowledge of the inner workings of our brains is still quite limited, despite intensifying ongoing efforts to deconstruct it, like the Human Brain Project in Europe and BRAIN in the United States.

That hasn't gotten in the way of making chips to structurally mimic the brain. This field, known as neuromorphic computing, began with the work of Carver Mead at Caltech in the 1980s, as he sought not to make better computers but to figure out "how in the heck can the brain do what it does?"[61] The reverse engineering of the brain into chips has included silicon neurons that use less energy and, like the brain, decentralize the architecture from single, do-it-all chips to simpler, dedicated ones that distribute the work and lessen the power. Neuromorphic chips have enhanced our understanding of brain circuitry and have paved the way for hardware systems for brain-machine interfaces and neuroprosthetics in the future. Indeed, AI has been used, in patients with deep brain implants for epilepsy, to develop individual models for how each person (of twenty-five individuals) remembers things. The implant electrodes sat quietly but were stimulated by trained algorithms sensing memory was needed and so provided a boost.[62] Perhaps the best demonstration of a fusion of AI and neuroscience is the work being done to create "biohybrid" computers by integrating silicon neurons with biological neural networks.[63]

More efforts are being expended by the chip industry to exploit knowledge of the brain's circuitry to design specialty chips. As John Hennessy, former president of Stanford University, put it, "The existing approach is out of steam, and people are trying to re-architect the system."[64] By off-loading the training of neural networks, which have hundreds of algorithms, onto such low-power dedicated chips, efficiency is improved, and computing power can be conserved.

Most of the AI in neuroscience work for neural networks has involved software and algorithm development. A mixed hardware-software approach for artificial synapses was undertaken by IBM

Research, creating a neural network with over 200,000 two-tiered (short- and long-term) synapses for image recognition, requiring one hundred times less power and, at the same time, creating efficiency of more than 28 billion operations per second per watt (which, compared with current graphic processing units, is more than two orders of magnitude greater). This accomplishment bodes well for future jumps in efficiency and reduced power requirements for artificial neural networks.[65]

So, as efforts to recapitulate the functioning of the human brain in computers—indeed, to make them more powerful—pushes forward, it brings us back to the question that opened this chapter: Will science ever be done by computers alone?

THE SCIENTIST'S NEW TOOLS AND APPRENTICE

Despite the clear power of AI to help scientists make new discoveries, it hasn't been universally adopted. Even as Xiaohui Xie lauded the work by Jian Zhou and Olga Troyanskaya, he acknowledged that "people are skeptical. But I think down the road more and more people will embrace deep learning." Clearly, I agree. But, for the doubters, there's much more to come that should convert them, as AI promises to revolutionize the practice of science, whether one is discussing the tools scientists use or the source of the ideas we test on them.

The microscope has been an iconic tool of biomedical scientists for centuries. It underwent one revolution with the development of a technique called fluorescence microscopy, which won the Nobel Prize for its inventors in 2014. Fluorescence microscopy involves complex sample preparation that attaches fluorescing molecules to cells, subcellular features, and molecules to make them visible through a microscope. Besides the time-consuming prep issue, this labeling harms and kills cells, which both makes specimens into artifacts and preempts the possibility of serial, longitudinal assessment of samples.[66] Enter deep learning. Eric Christiansen and colleagues at Google, with collaborators at the Gladstone Institute and Harvard, developed open-source algorithms that can accurately predict how samples would fluoresce without the need for any fluorescent preparation. They trained the DNN by matching fluorescent labeled images with unlabeled ones, repeating the process millions of times. This method, known as in silico labeling as well as augmented microscopy was called "a new epoch in cell biology."[67] Their assertion was quickly followed by another report of label-free microscopy by scientists from the Allen Institute.[68] Beyond whole cells, accurate classification of subcellular images at scale was achieved by two different approaches—both deep learning models and involving more than 320,000 citizen scientists. The complementarity of machine learning and human brain processing in achieving high levels of accuracy was noteworthy.[69]

Likewise, machine learning has led to "ghost cytometry." It can be extremely difficult to identify, sort, or capture rare cells in the blood. Researchers at ThinkCyte in Japan developed algorithms that detect cell motion to enable highly sensitive, accurate, ultrafast sorting of cells that did not require image production.[70] Similarly, the University of Tokyo led the deep neural network development of "intelligent image-activated cell sorting"[71] for real-time sorting of various types of cells.

Beyond these image-free, label-free breakthroughs, deep learning for microscopy has been shown to help deal with suboptimal, out-of-focus images[72] and to accelerate super-resolution, reconstructing high-quality images from undersampled light microscopy data.[73] It has also been used to detect metastatic cancer in real time, expediting the readout of pathology slides.[74]

As radical a change as those developments in microscopy are, they pale beside the plans some researchers have for the automation of science. Here machines aren't just running (battery testing, chemical reagent) experiments, they're getting ready to devise them. The concept of fully automated science and a full-fledged machine colleague seems as far off to me as it does alien. But when I saw the *Bloomberg* article subtitle "Carnegie Mellon Professors Plan to

Gradually Outsource Their Chemical Work to AI," I wondered just how far off it might actually be.[75] I got the sense that we're taking steps toward it when I read about machines developed by Zymergen, one of many companies that are working to change the current repertoire of lab robots, remarkably quiet like crickets singing and barely noticeable as they did their work:

> Instead of using a pipette to suck up and squirt microliters of liquid into each well—a tidal wave of volume on the cellular scale—the robot never touches it. Instead, 500 times per second, a pulse of sound waves causes the liquid itself to ripple and launch a droplet a thousand times smaller than one a human can transfer.[76]

Automating a mechanical function of scientists is a well-accepted tactic. But AI promises to do more. There are many science apprentice functions that AI can help with, including conducting far better literature searches (as with Iris.ai and Semantic Scholar), designing or running experiments (as with Zymergen and Transcriptic), interpreting data (like Nutonian, which produces a mathematical theory based in data ingestion), and writing the paper (with Citeomatic, which finds missing citations in the manuscript draft).[77] In cellular and molecular biology, the manual labor of plating cells and counting colonies can be preempted. Accuracy and efficiency of executing certain experiments have been enhanced. Some researchers have embraced AI for its data-driven approach to "designing" (many have questioned this term since it involves human intuition) the next set of experiments. The concept of "accelerating the scientific method" has already been validated by many of the advances I've summarized and so many more in the pipeline.[78] But it's fair to say there are plenty of constraints for the types of lab-related work that can even be partially automated by AI tools.

Apprenticeship opportunities will continue to emerge throughout all disciplines of science. The areas we've covered here—neuroscience, cancer, omics, and drug discovery—represent the leading edges just as the doctors with patterns (radiologists and pathologists) are for medicine. The parallels in science of increasing efficiency and seeing things that humans can't but machines can are striking. I don't believe we will ever progress to "ghost" scientists, replaced by AI agents, but off-loading many of the tasks to machines, facilitating scientists doing science, will, in itself, catalyze the field. It's the same theme as with doctors, acknowledging we can develop the software that writes software—which in turn empowers both humans and machines to a higher order of productivity, a powerful synergy for advancing biomedicine.

Now let's shift our attention away from what AI can do for doctors, health systems, and scientists to what it can do for consumers. To start that off, one of the most important yet controversial and unresolved aspects of our health—our diet.

chapter eleven

DEEP DIET

> The drug we all take multiple times a day that needs to be personalized most is food.
>
> —Lisa Pettigrew

AFTER MY SECOND BOUT WITH KIDNEY STONES, MY UROLO-gist advised me that it was essential that I see his nutritionist. That appointment took weeks to schedule. In the meantime, with the results from my twenty-four-hour urine test in hand, I read up on diet modifications that could help to reduce my high urine oxalate level (mine was 64, and the normal range is from 20 to 40). Looking at multiple websites and articles, I was struck by the remarkable inconsistency of oxalate quantities in various foods—depending on what I read, the same foods could either be deemed okay or out of the question. Litholink, the testing company that did my urine analysis, had very high values for Fiber One cereal (142 mg oxalate/100 g), black pepper (419), chocolate (117), and spinach (600) but quite low values for sweet potatoes (6), kale (13), and blueberries (15). But a University of Pittsburgh Medical Center website for low oxalate diet listed blueberries, sweet potatoes, and

	BLUEBERRIES	STRAWBERRIES	KALE	SWEET POTATOES
Litholink	Very low	Not available	Low	Low
University of Pittsburgh Medical Center	High	Not available	High	High
Academy of Nutrition and Dietetics	High	High	High	High
Harvard School of Public Health	Very low	Very low	Very low	Very high

TABLE 11.1: Comparison of oxalate content for four foods by four sources.

kale as high in oxalate. These were just a few examples of striking variance in recommendations between the two resources. Unsurprisingly, I was confused, and I was looking forward to meeting the nutritionist to clear everything up.

My dietician had more than twenty years of experience as a professional nutrition specialist. She reviewed my lab test data and then reviewed her recommendations for what I should eat, supported by a three-page document from the Academy of Nutrition and Dietetics. She advised me to eliminate nuts and spinach, two of my favorite foods, avoid strawberries and blueberries, more of my favorite foods. But Litholink said all of these were low in oxalates. Now I was even more confused. After the visit, I reviewed the prior materials and other websites, and e-mailed her for clarification along with a request for the best source of data. She wrote back that she had recommended avoiding strawberries because they are so large that it's quite easy to eat more than the half-cup recommended serving, increasing my oxalate intake. She also said that, because different sources will list the serving sizes of fruit in mass, by ounces or grams, or by volume, it was very easy for different sites to categorize the same food as low, moderate, or high.

She sent me a link to the source of her recommendations, the department of nutrition at the Harvard T. H. Chan School of Public

Health, to review. Under fruits, it reports that a half cup of blueberries had very low oxalate, at 2 milligrams. Likewise, a half cup of strawberries had 2 milligrams of oxalate. On the extreme, a cup of raspberries was very high, at 48 milligrams. Under vegetables, kale 1 cup chopped, very low, 2 milligrams. In contrast spinach, raw, 1 cup, was very high, at 666 milligrams; sweet potatoes 1 cup, very high at 28 milligrams (Table 11.1). You get the picture. Everything remarkably inconsistent from one source to another. What and whom do I believe?

My experience is emblematic of the state of nutritional science. Hippocrates, sometime back around 400 BC said, "Let food be thy medicine and medicine be thy food." Although we've thought that our diet and health are interconnected for millennia, the field is messy. One major problem is that it's exceptionally hard to do any randomized trials at scale. Assigning a large cohort of people to one diet or another that requires close adherence over many years, and then following major health outcomes, is very difficult and rarely attempted. The primary exceptions are the randomized trials done of the Mediterranean diet, showing a 1 to 2 percent absolute reduction in heart disease.[1] But even the largest diet randomized trial—the PREDIMED Mediterranean diet—had to be redacted and republished, the subject of methodological issues and analytical controversy following charges of flawed statistics.[2]

Most nutrition science is predicated on observational and retrospectively gathered data, which is dependent on people accurately reporting what they eat. "Accurately reporting" could be thought of as an oxymoron. John Ioannidis, a respected critic of scientific methodology, tore apart current nutritional science analytical methods,[3] as did Bart Penders.[4]

Nevertheless, let's review a few of the recent large observational studies that looked at diet and major outcomes. The Prospective Urban Rural Epidemiology (PURE) paper, published in the *Lancet* in 2017, ranked number one for the year by Altmetric (168 news stories, 8,313 tweets, 441 Facebook posts), studied more than 135,000 people from eighteen countries. It identified high carbohydrates,

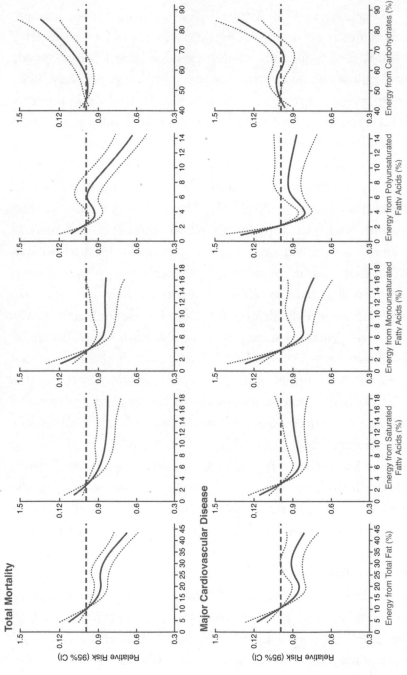

FIGURE 11.1: The PURE study association between estimated percent nutrients and all-cause (total mortality) and major cardiovascular disease. The dotted lines represent the 95 percent confidence intervals. Source: Adapted from M. Dehghan et al., "Associations of Fats and Carbohydrate Intake with Cardiovascular Disease and Mortality in 18 Countries from Five Continents (PURE): A Prospective Cohort Study," *Lancet* (2017): 390(10107), 2050–2062.

and not fat intake, as the key culprit in the risk of heart disease and death (Figure 11.1).[5]

Another 2017 study in the United States examined the consumption of ten foods and nutrients in more than 700,000 people who died from heart disease, stroke, or diabetes.[6] For example, diets high in salt or processed meats, as well as those low in seafood or fruits and vegetables, were all implicated in significant bad outcomes (Figure 11.2). The conclusion of the study was that 45 percent of these deaths can be attributed to these ten factors with "probable or convincing evidence." If true, that means that about one in two deaths from heart disease, stroke, or diabetes is linked to poor diet. All told, that means more than a thousand Americans die as a result of their diets each day.

Other studies have suggested that plant-based diets can help prevent type 2 diabetes.[7] Besides plant-based foods, whole grain consumption was associated with less heart disease or cancer deaths when the results of forty-five studies were pooled.[8] And there's no shortage of data associating coffee and improved survival.[9] Each of these studies is compromised by several characteristics: they rely on self-reporting of nutritional intake; they have no capacity for demonstrating cause and effect; they have no controls; and they were not constructed to sort out a surfeit of potentially confounding factors, including socioeconomic status and educational attainment. Indeed, a systematic review by Jonathan Schoenfeld and John Ioannidis showed association for cancer for both risks and benefits of most foods.[10] The media reporting to the public typically doesn't take these critical caveats into account, so we have a blitzkrieg of erroneous, misguided headlines, with every type of food helping or harming depending on the day and outlet (Figure 11.3).

The lack of gold-standard randomized clinical trials is only one dimension of the problem facing the science of human nutrition. One major issue has been the effect of poorly done research in the formulation of dietary advice. In *The Big Fat Surprise,* Nina Teicholz,

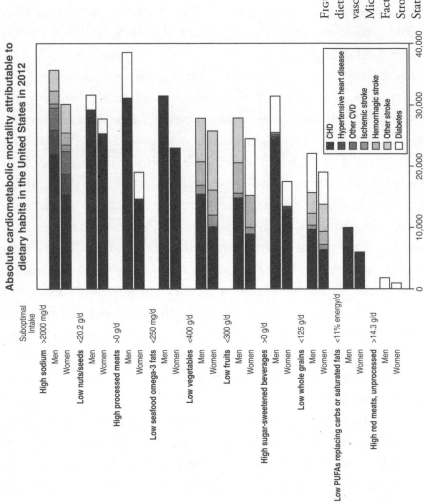

FIGURE 11.2: Association of suboptimal dietary habits with major adverse heart and vascular outcomes. Source: Adapted from R. Micha et al., "Association Between Dietary Factors and Mortality from Heart Disease, Stroke, and Type 2 Diabetes in the United States," *JAMA* (2017): 317(9), 912–924.

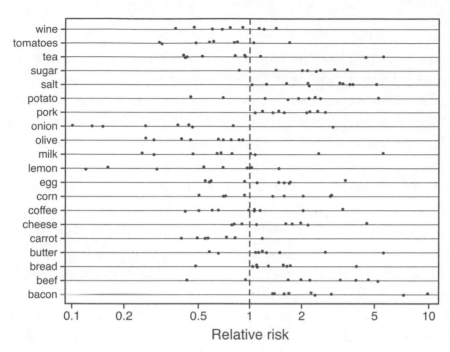

FIGURE 11.3: Effect estimates in the literature for specific foods and risk of cancer. Source: Adapted from J. Schoenfeld and J. Ioannidis, "Is Everything We Eat Associated with Cancer? A Systematic Cookbook Review," *Am J Clin Nutr* (2013): 97(1), 127–134.

an investigative journalist, chronicled the impact of physiologist Ancel Keys on what we should eat. He had published research on what he called the seven countries diet and appeared in a cover story in *Time* magazine in 1961, advocating a low-fat, low-cholesterol diet to prevent heart disease. But Keys's study was flawed, omitting fifteen countries from which he had gathered contradictory data, leading to criticism of its findings at the time. Nonetheless, the American Heart Association strongly promoted a low-fat diet, which included advocating margarine instead of butter and avoidance of eggs. For years, like many of you, I did everything possible to avoid fat, living on stuff like nonfat pretzels and black coffee, and running away from cheese and pizza. Even 1 percent milk was off-limits. The contrast to the prior era when I was growing up in the late 1950s and early 1960s was striking: my family had several bottles of whole milk delivered to our house a few times a week, and my parents called me "Elsie the cow" because I drank so much of it (even though that didn't make sense). Only decades later did we become aware of the toxic heart effects of margarine, stuffed with trans fats, which were eventually banned as food ingredients in many countries. Still the American Heart Association and US Department of Agriculture recommend limiting saturated fat in their guidelines. Both aspects of the story exemplify the misdirected promotion of diet advice without adequate data. The result has been that, by endorsing low-fat foods, health agencies were likely endorsing a harmful diet, fueling the obesity and diabetes epidemic. In fact, the long-standing recommendations for avoidance of dairy products and salt intake have been seriously challenged by recent reports.[11]

The next issue is corruption in the food industry, exemplified by the sugar scandal. Sugar is in three-fourths of all packaged foods.[12] Since the 1950s, the sugar industry has promoted the idea that a calorie is a calorie, and that eating a calorie's worth of sweets is no more likely to make someone obese than any other food.[13] The sugar trade association put the blame for heart disease on saturated fat. For decades, the sugar industry has been commissioning researchers, including the influential Ancel Keys, to echo these claims. The three Harvard scientists who published the classic review paper indicting dietary fats for heart disease in the *New England Journal of Medicine* in 1967 were paid by the Sugar Association.[14] The trade association also opposed new labels on food that would disclose how much sugar had been added to packaged foods. That problem continues in the present day. In 2015, we learned that Coca-Cola worked with scientists to squash the notion that sugar had anything to do with obesity. This is not a problem only with the sugar industry. Marion Nestle has shown that, in nearly two hundred food studies, those funded by industry (compared to no industry support) had a positive ratio of thirteen to one.[15] Food "science" is not only compromised by a dearth of hard evidence but also colored by bias.

The confused state of dietary recommendations is furthered by government-issued food pyramids—pyramids that are built on soft ground, even though many agencies are involved, including the USDA, NIH, CDC, FDA, and EPA. Because it's the government publishing the information, it's easy to accept the information as gospel. Maybe we can extend the adage "too many cooks spoil the broth" to "too many agencies have spoiled the guidelines."

For decades we've been led to believe that too much salt in the diet dangerously raises the risk of heart attacks and strokes. The American Heart Association still recommends no more than 1.5 grams of sodium per day. If you've ever tried eating such a low-salt diet, you've likely found it unpalatable, even intolerable. Maybe it would be good for weight loss since foods become tasteless. Sir George Pickering once said, "To stay on it requires the asceticism of a religious zealot."[16] But the link of excess sodium with risk for adverse heart and vascular events has been debunked. A 2018 study of more than 95,000 people in eighteen countries, while verifying a modest increase of blood pressure (with increasing sodium ingested as reflected by urine measurement), showed that the bad outcomes only occurred when sodium intake exceeded 5 grams per day.[17] The average American takes in about 3.5 grams of sodium per day.[18] In fact, for less than 5 grams per day of sodium, there was an inverse correlation between sodium intake and heart attack and death! One more example of how we have long-lived national, nutritional recommendations that don't stand up to evidence. And how we are stuck on averages instead of recognizing or understanding the marked individuality of food responsiveness.

This indeed is the biggest problem facing nutrition guidelines—the idea that there is simply one diet that all human beings should follow. The idea is both biologically and physiologically implausible, contradicting our uniqueness, the remarkable heterogeneity and individuality of our metabolism, microbiome, environment, to name a few dimensions. We now know, from the seminal work of researchers at Israel's Weizmann Institute of Science, that each individual reacts differently to the same foods and to precisely the same amount of a food consumed. The field of nutrigenomics was supposed to reveal how our unique DNA interacts with specific foods. To date, however, there is very little to show for the idea that genomic variations can guide us to an individualized diet—the data are somewhere between nonexistent and very slim. That hasn't stopped many companies from marketing the concept. Nutrigenomics companies are marketing assays of particular DNA sequence variants to shape your diet, although they have little to no grounding in acceptable evidence[19] or have even been debunked by randomized trials.[20] Indeed, the veracity of many food science sources has been called into question.[21] Likewise, the use of smartphone apps to make recommendations for food by companies providing a virtual nutritionist function include Suggestic, Nutrino, and Lose It!, but the basis, the science, for their individualized guidance is not clear. To transcend the evidence-free, universal diet concept, it takes a computational, data-driven, unbiased approach. And that is where artificial intelligence came into play. Indeed, the Weizmann research didn't just show that different people will experience different outcomes from eating the same foods. For the first time, machine learning played a critical role in understanding the issue, predicting each person's unique glycemic response to food.

In November 2015, the journal *Cell* published a landmark paper titled "Personalized Nutrition by Prediction of Glycemic Responses," authored by Eran Segal, Eran Elinav, and their colleagues at Weizmann.[22] The study included eight hundred people without diabetes, monitoring their blood glucose via a subcutaneous sensor for a week. In total, the participants were monitored over a period in which they ate more than 5,000 standardized meals, some of which contained items like chocolate and ice cream, provided by the researchers, as well as 47,000 meals that consisted of their usual food intake. In total, there were more than 1.5 million glucose measurements made.

The granular glycemic response to food and other stimuli was integrated with much more multidimensional data for each person: dietary habits such as time of meal, food and beverage content, phys-

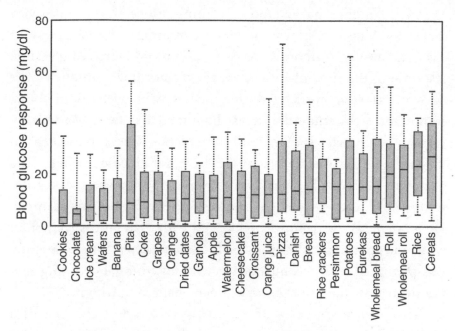

FIGURE 11.4: The average blood glucose (with the 25, 75th percentiles) increases in response to different foods in the Weizmann Institute studies. Note the highly variable responses to all foods, especially breads (pita, whole wheat), pizza, orange juice, potatoes. Source: Adapted from E. Segal and E. Elinav, *The Personalized Diet: The Pioneering Program to Lose Weight and Prevent Disease* (New York: Grand Central Life & Style, 2017).

ical activity, height and weight, sleep, gut microbiome, and blood tests. Much of these data were entered via a dedicated smartphone app by the participant. The post-meal glucose response to food was, as expected, highly variable (Figure 11.4).[23]

A decision-tree machine learning model crunched these millions of data points. It highlighted 137 factors that were used to predict the glycemic response to specific foods for each individual. This was validated in another one-hundred-person cohort. Then, providing yet another layer of confirmation of the algorithm's value, a randomized trial was done in twenty-six people with personalized diet plans and showed significantly improved post-meal glucose response (derived from machine learning) as compared with a control group. The algorithm was remarkably accurate in predicting the glycemic response and outperformed the predictions of expert nutritionists.

Such findings have important implications: for people with diabetes taking insulin, carbohydrate counting is the main way to calculate dosing. Carbohydrates were tied to an increased glycemic response at the time of eating, as was dietary fiber, although fiber decreased the response later, during the following twenty-four hours. Importantly, the study didn't just highlight that there were highly variable individual responses to the same food—it was able to explain it. The food constituents weren't the driver for glucose response. The bacterial species in the gut microbiome proved to be the key determinant of each person's glucose response to eating. For example, *Parabacteroides distasonis* was associated with high glucose response, while the opposite was the case for *Bacteroides dorei*. When the study was published, *Cell* accompanied it with an editorial declaring that the study was "the first step toward personalized nutrition."[24]

This Weizmann Institute group paper was just the first of a series of their publications. As a next step, they investigated modulating bread consumption as a means to intervene in blood glucose levels. Worldwide, bread provides about 10 percent of the calories people consume—in some parts of the world, bread provides even more than 30 percent—so they focused on this food as the intervention. In 2017, they reported a randomized, crossover study of two different bread types: industrial white or artisanal sourdough.[25] Each of the twenty persons in the study had continuous glucose monitoring and pretty much the same protocol of data collection as in the first study. The description of the bread preparation is colorful, shows how much detail these researchers get into, and makes me hungry:

> For white bread, we provided people in the study with a standard popular white bread, to be sure everyone was eating the same bread. To create the sourdough bread, we hired an experienced miller to stone mill fresh hard red wheat and sift the flour to remove only the largest bran particles. We also hired an experienced artisanal baker to prepare loaves of bread using only specially milled flour,

water, salt, and a mature sourdough starter, without any other additives. The dough was portioned and shaped, proofed, and baked in a stone hearth oven. Every two days, we brought this freshly baked whole-grain sourdough bread to our labs to give to the study participants. The smell was so enticing that it was difficult to keep our team members away! Knowing a losing battle when we saw one, after the second delivery, we started ordering extra loaves for our lab members.

The results were very surprising. There was no difference in overall results for the glycemic response to different breads. But that was because they were looking at population averages. At the individual level, there was striking variability. Some people had a low glycemic response to white bread and others had just the opposite. Again, the gut microbiome was the driver. In fact, in the case of these two bread types, it was not just the driver but the sole predictor.[26]

Our individual gut microbiomes—approximately 40 million communal cells of a thousand different species—play a far bigger role in response to food intake than we had anticipated. There have been many studies that tie the gut microbiome to issues related to diet, including obesity and diabetes, as well as immune disorders and a long list of other conditions, but without unequivocal proof of a cause-and-effect relationship. This might be because we shed about 10 percent of our gut microbiome in our stool each day—perhaps the populations are just too variable to have a reliable effect. However, the overall diversity of species and content tends to remain the same. There are other factors that do influence the composition of the microbiome. Of note, these bacteria have their own circadian rhythm, some being more abundant in the morning or evening. That rhythm is controlled by both our eating patterns and our own biological clocks. For example, the Weizmann group did a study in which they gave participants free round-trip tickets from Israel to the United States. They transferred gut microbes from the study participants, who were suffering from maximal jet lag, into germ-free mice.

That induced obesity and glucose intolerance.[27] In separate studies, the Weizmann team demonstrated that the deleterious effects, including weight gain and obesity, of ingesting artificial sweeteners[28] were correlated with changes in the microbiome.[29]

The extensive body of work done by Segal and Elinav is summarized in their book *The Personalized Diet.* Cumulatively, they've studied more than 2,000 people and summed up their revelations about nutrition science as "we realized we had stumbled across a shocking realization: *Everything was personal.*"[30] To quote a key conclusion in their book: "Because our data set was so large and our analysis so comprehensive, these results have an enormous impact—they show more conclusively than has ever been shown before that a generic, universal approach to nutrition simply cannot work." That's the kind of bold statement you wouldn't find in a peer-reviewed journal article, but the kind of strong assertion you might find in a book.

This individuality was related to glycemic response, which is an important but certainly not the end-all metric of nutrition impact and human health. Glycemic spikes after eating, particularly when they are substantial, could turn out to be a harbinger of increased risk for developing diabetes,[31] and high glucose has been mechanistically linked to permeability of the gut lining, increasing the risk of infection[32] and cancer.[33] Beyond the potential link to diabetes and cancer, concerns have always been registered for blood lipid abnormalities, obesity, and heart and neurodegenerative diseases. But, for now, no loop has been closed between healthy people's glucose spikes and diseases.

Without question, these researchers have shown that individualized patterns for glycemic response—some people were very fat sensitive, others fiber responsive, some sodium sensitive, others very affected by their sleep—were tied to the gut microbiome, and that the complexity could be mapped, modeled, and predicted via machine learning algorithms. Subsequently, a group at Stanford assessed the glycemic spikes in fifty-seven healthy people with continuous glucose monitoring, analyzed the response to specific food data with machine

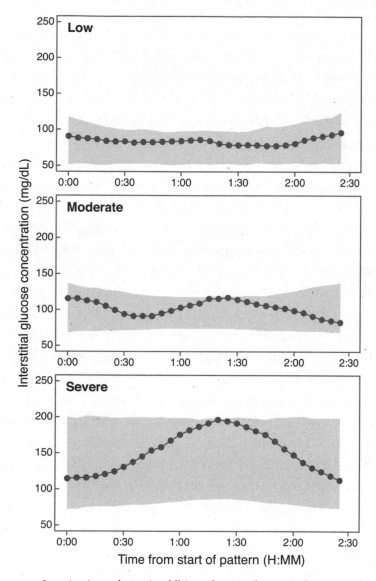

FIGURE 11.5: Low (top), moderate (middle), and severe (bottom) glucotypes after eating in fifty-seven healthy individuals. Source: Adapted from H. Hall et al., "Glucotypes Reveal New Patterns of Glucose Dysregulation," *PLoS Biol* (2018): 16(7), e2005143.

learning, and suggested the glucose spikes after eating are common and fall into three "glucotypes" (Figure 11.5).[34] Certain foods were particularly implicated: "A standardized meal of cornflakes and milk caused glucose elevation in the prediabetic range (>140 mg/dl) in 80 percent of individuals in our study. It is plausible that these commonly eaten foods might be adverse for the health of the majority of adults in the world population." Such findings of both the Weizmann and Stanford reports, for glucose spikes and implicating the gut microbiome, have been confirmed by others.[35]

Segal and Elinav, knowing my interest in individualized medicine, asked me to review their new book while it was in the galley proof stage—it clearly summarized an important body of research, providing cogent evidence for our variable blood glucose levels in response to food. They had also decided to form a company in Israel in 2015 to help people determine an individualized, optimal diet with respect to glycemic response, DayTwo with Lihi Segal, Eran Segal's wife, at the helm as CEO. Although I was already intrigued by their scientific publications, reading the book prompted me to ask whether I could try DayTwo. I began with filling out a web survey to provide my demographics, downloading the DayTwo health app, and receiving the Libre glucose sensor from Abbott Laboratories. Once I was fully on board, I logged everything I ate and drank, my sleep data, exercise, and medications via my smartphone for two weeks. I wore the sensor, about the size of a half dollar, on my left arm throughout that time. I could quickly check my glucose at any time with the dedicated reader that came with it. I also had to collect a stool specimen for my gut microbiome assessment.

This two-week data collection was certainly a hassle. The glucose and microbiome parts were easy, and I was able to use my Fitbit to export data on sleep and activity, but manually entering everything that I ate or drank on my phone was tedious. Finding the food or beverage from the list of choices, along with the portion size, was often inexact. Not infrequently, I had to go back a day or two previously to fill in the information since I had gotten busy or forgot. I was instructed to not eat anything for at least two hours after a meal so that the glycemic response wouldn't get conflated. That was difficult at times since I might otherwise have had a little snack. The rule reminded me of a fascinating article from the Salk Institute by

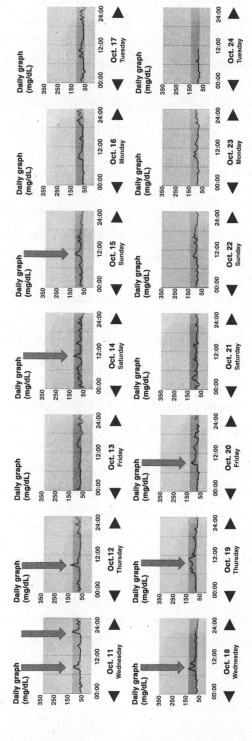

FIGURE 11.6: My glucose monitoring for two weeks showing multiple spikes (arrows) as high as 150 mg/dl after eating.

Satchin Panda, who used a smartphone app to monitor daily eating patterns. Panda's work showed people completely lack a three-meals-a-day structure but instead eat for a median of 14.75 hours a day![36] The Hawthorne effect—whereby people in a study behave differently when they know they are being watched—was certainly coming into play, perhaps causing me or any other participant in DayTwo's program to eat or avoid certain foods or otherwise change our dietary habits.

With those caveats, here are my results of my glucose, my microbiome, and my food recommendations. I had several glucose spikes after eating (Figure 11.6), fitting into the moderate glucotype.

Notwithstanding the bother, the information on my microbiome (Figure 11.7) and food recommendations was very interesting. One bugger in particular—*Bacteroides stercoris*—seems to be my special coinhabitant. The food recommendations (Figure 11.8) suggest I am very sensitive to carbohydrates, but not so much to fats, for my glycemic response. The list of foods in each category provides a far broader menu of choices than what I ate during the data collection period. From its algorithm, DayTwo also suggested meals that would keep me in a tighter glycemic range. Plus, it has a database of more than 100,000 foods that I can now search to see whether any are predicted to cause a glycemic spike for me.

Since I did this data collection, DayTwo has changed its strategy. While initially in Israel it did the glucose monitoring and extensive self-tracking, its US launch involved only taking a gut microbiome sample and having the algorithm predict optimal dietary choices for $329. Subsequently, DayTwo adopted the same plan in Israel, so what I experienced is no longer an offering. DayTwo is not the only company in this space: Viome is a competitor, assessing the microbiome more comprehensively (not only bacteria but also viruses and fungi) for $399 and then using the data to recommend an individualized diet.[37] Unlike the Weizmann Institute's serial reports, however, Viome has not published any peer-reviewed research so far.

Microbiome

Item	Percentage
Bacteroides stercoris	You: 27.45% / Population Ave: 1.74%
Bacteroides vulgatus	You: 9.37% / Population Ave: 2.49%
Bacteroides uniformis	You: 9.25% / Population Ave: 2.75%
Eubacterium rectale	You: 5.96% / Population Ave: 4.81%
Alistipes putredinis	You: 5.62% / Population Ave: 3.16%
Clostridium sp. L2-50	You: 4.13% / Population Ave: 0.84%
Faecalibacterium prausnitzii	You: 4.09% / Population Ave: 6.80%
Ruminococcus bromii	You: 3.90% / Population Ave: 3.10%
Parabacteroides merdae	You: 3.49% / Population Ave: 1.33%
Barnesiella intestinihominis	You: 3.31% / Population Ave: 1.36%
Bacteroides ovatus	You: 2.46% / Population Ave: 0.98%
Bacteroides cellulosilyticus	You: 1.83% / Population Ave: 0.61%
Roseburia intestinalis	You: 1.38% / Population Ave: 1.06%
Bacteroides dorei	You: 1.22% / Population Ave: 1.53%
Bacteroides faecis	You: 1.10% / Population Ave: 0.33%
Anaerostipes onderdonkii	You: 1.02% / Population Ave: 0.57%
Akkermansia muciniphila	You: 0.88% / Population Ave: 1.50%
Bifidobacterium adolescentis	You: 0.87% / Population Ave: 2.57%
Acidaminococcus sp. D21	You: 0.86% / Population Ave: 0.06%
Parabacteroides distasonis	You: 0.85% / Population Ave: 0.53%
Ruminococcus lactaris	You: 0.77% / Population Ave: 1.59%
Dorea longicatena	You: 0.73% / Population Ave: 1.53%
Eubacterium hallii	You: 0.69% / Population Ave: 0.09%
Odoribacter laneus	You: 0.44% / Population Ave: 0.23%
Lachnospiraceae bacterium 3157FAA	You: 0.44% / Population Ave: 0.44%
Roseburia hominis	You: 0.41% / Population Ave: 1.01%
Alistipes shahii	You: 0.39% / Population Ave: 1.16%
Ruminococcus torques	You: 0.39% / Population Ave: 0.84%
Ruminococcus obeum	You: 0.37% / Population Ave: 0.27%
Eubacterium ventriosum	You: 0.36% / Population Ave: 0.87%
Eubacterium biforme	You: 0.32% / Population Ave: 0.79%
Coprococcus comes	You: 0.29% / Population Ave: 0.52%
Lachnospiraceae bacterium 1157 FAA	You: 0.28% / Population Ave: 0.07%
Bacteroides fragilis	You: 0.26% / Population Ave: 0.54%
Dialister invisus	You: 0.23% / Population Ave: 1.31%
Roseburia inulinivorans	You: 0.15% / Population Ave: 0.39%
Eubacterium eligens	You: 0.14% / Population Ave: 1.22%
Coprococcus catus	You: 0.12% / Population Ave: 0.51%
Bifidobacterium longum	You: 0.12% / Population Ave: 0.27%
Dorea formicigenerans	You: 0.11% / Population Ave: 0.98%
Ruminococcus callidus	You: 0.08% / Population Ave: 0.12%
Collinsella aerofaciens	You: 0.07% / Population Ave: 0.40%
Haemophilus parainfluenzae	You: 0.06% / Population Ave: 0.08%
Alistipes finegoldii	You: 0.05% / Population Ave: 0.46%
Alistipes senegalensis	You: 0.04% / Population Ave: 0.24%
Bacteroides thetaiotaomicron	You: 0.03% / Population Ave: 0.08%
Streptococcus thermophilus	You: 0.03% / Population Ave: 0.03%
Clostridium miele	You: 0.03% / Population Ave: 0.37%
Bilophila wadsworthia	
Lachnospiraceae bacterium 5163 FAA	

FIGURE 11.7: My gut microbiome assessment, with my predominant *Bacteroides stercoris* coinhabitant.

Bread, Cereal, Rice, and Pasta

Item	Grade
French toast challah bread	A
Granola	A-
Crackers	B+
Baguette with Camembert	B
Oatmeal with milk	B
Whole wheat bread with butter	B
Quinoa	B-
Bran flakes with soy milk	B-
Gluten free whole grain bread w/olive oil	B-
Corn tortilla	C+
Multigrain crackers	C+
Oatmeal	C+
Cooked buckwheat	C
Banana nut crunch cereal with milk	C
Ciabatta bread with avocado	C
Cheetos	C-
Gluten-free bread	C-
Italian-herb focaccia	C-
Multigrain rice cakes	C-
Salted rice cakes	C-

Vegetables

Item	Grade
Cooked broccoli	A+
Cooked cauliflower	A+
Kimchi	A+
Yellow beans	A+
Prepared cauliflower	A+
Artichokes	A
Beets	B+
Winter squash	B
Cooked brussel sprouts	B-
Baked sweet potatoes	B-
Lima beans	B-
Yams	C+
Baked squash	C
Yellow peppers	C
Baked potatoes	C-
Celeriac	C-.
Pickled daikon radish	C-

Beverages

Item	Grade
Decaf instant coffee	A+
Light beer	A+
Martini	A+
Cappuccino	A
Pina colada	A
American-style pale lager	A-
Sweetened vanilla coffee	A-
Cola	B-
Cranberry juice	B-
Orange juice	B-
Fruit punch	C+
Guava passion fruit juice	C+
Spicy apple cider	C+

Legumes, Tofu, and Nuts

Item	Grade
Almond butter	A+
Brazil nuts	A+
Mixed nuts	A+
Sunflower seeds	A+
Tahini spread	A+
Edamame	A
Unsalted trail mix	A
Nuts, seeds & raisins trail mix	A-
Roasted & salted pumpkin seeds	B+
Berry blend trail mix	B+
Home prepared hummus	B
Soy burger	C+
Spicy black bean burger	C+
Roasted chestnuts	C
Lentil veggie burger	C
Veggie burger	C

Fruits

Item	Grade
Star fruit	A+
Strawberries	A+
Unsweetened coconut	A+
Blackberries	A
Asian pear	A-
Guava	A-
Raspberries	A-
Nectarine	B-
Pear	B-
Plum	B-
Pomegranate	B-
Tangerine	B-
Banana	C+
Cherries	C+
Dried cherries	C+
Goji berries	C+
Orange	C+
Melon	C
Raisins	C
White grapefruit	C
Dried papaya	C-
Pomelo	C-

Meat, Fish, and Eggs

Item	Grade
Bratwurst veal cooked	A+
Hard-boiled egg	A+
Smoked salmon	A+
Breaded veal cutlet	A+
Fried mullet	A+
Grilled chicken breast	A
Spicy shrimp ceviche	A
Cod cakes	A
Pickled Atlantic herring	A-
Salmon sashimi	A-
Fried squid	B+
Fish sticks	C-

Dairy and Dairy Substitutes

Item	Grade
Almond milk	A+
Blue cheese	A+
Goat's milk	A+
Gouda cheese	A+
Soy cheddar cheese	A+
Plain whole yogurt	A
Whole milk	A
Greek yogurt	B+
Soy milk	B+
Berry soy yogurt	B-
Skim milk	B-
Soy yogurt	B-
Berry yogurt 0% fat	C
Chocolate soy milk	C-
Yogurt 0% fat	C-

Snacks and Sweets

Item	Grade
Cheese danish	A
Cheesecake	A
Fiber almond brownie bar	A-
Protein almond brownie bar	A-
Carrot cake with extra icing	B+
Raspberry white chocolate muffin	B+
Chocolate cream sandwich cookie	B+
Almond raisin cinnamon danish	B
Apple, cinnamon, raisin & strawberry danish	B
Coffee cake	B
Raspberry danish	B
White chocolate macadamia cookies	B
Mini chocolate chip muffin	B-
Peach pie	B-
Pecan pie	B-
Fruit & hazelnut bar	C+
Ice cream sandwich	C+
Strawberry frozen yogurt	C+
Apple cinnamon muffin	C
Chocolate cake cookie	C
Honey graham crackers	C
Banana nut muffin	C-
Wheat bran raisin muffin	C-
Whole wheat fig bars	C-

FIGURE 11.8: Individualized recommendations with grades for me based upon the DayTwo algorithm.

The work from Elinav and Segal's lab is not the only one that has reinforced the status of the microbiome as pivotal to each individual's response to intake. Michael Snyder, who heads up the genetics department at Stanford University, led a multi-omic study (assessing microbiome, transcriptome, proteome, metabolome, and genome) in twenty-three overweight individuals to characterize what happens with weight gain and loss. With as little as a six-pound weight gain, there were dramatic changes in the gut microbiome species, more than 300 genes exhibited significant change in function, and a release of pro-inflammatory mediators showed up in the blood.[38] And these substantial changes were fully reversed with weight loss.

Let me be very clear: I don't present my findings with DayTwo to recommend this company or idea. It's clearly interesting, since it represents one of the first uses of AI for consumer health. It's also notable because it integrated so much data about individuals, something we haven't seen much of to date. But it's hardly been proven to make a difference. That would require a large randomized trial with half the people using the algorithm and half without that guidance, and then following up over years to see differences in clinical outcomes. All we have right now is a short-term story of one aspect of nutrition—the glucose response to food. That's very different, for example, from preventing diabetes or the complications that can result from diabetes. I'm also concerned that, without daily tracking of diet, activity, and glucose, the predictive algorithm will be less effective. When I asked the company about whether the ROC curves were different when the microbiome alone was assessed instead of the full package, I didn't get a response except that the microbiome itself was quite accurate. And Rob Knight, a top-flight microbiome expert at UC San Diego, said the Weizmann work is "very solid and rigorous, which puts them ahead of the curve. But I think it is still very challenging to extend results like that beyond the population you have studied directly."[39]

There's another problem here, which takes me back to the opening of this chapter. Since I have calcium oxalate kidney stones related to high oxalate in my urine, I'm supposed to be following a low-oxalate diet. You'll recall some of my favorite foods that I'm supposed to avoid, at least according to some sources, but that are rated A+ on DayTwo recommendations. The clash between general dietary recommendations, provided without knowledge of my metabolic disorder, and the specific diet plan for the latter illustrates the complexity here—the need to factor in all of a person's data to approximate a true, personalized diet. New technology, like an ingestible electronic capsule that monitors our gut microbiome by sensing different gases, may someday prove useful for one key dimension of data input.[40] We've seen bacteria genetically engineered to treat a metabolic disease via the gut microbiome (in a primate model).[41] For the time being, however, we remain a long way from achieving scientifically validated individualized eating, but this path will likely lead us to a better outcome than the one we've been stuck with, relying on universal diet recommendations.

Setting the table with our diet, we've got the perfect launch for the next chapter, which is how AI—well beyond the ability to inform an individualized, tailored diet—will promote consumer health and how our virtual assistant can take on medical coaching responsibilities.

chapter twelve

THE VIRTUAL MEDICAL ASSISTANT

Today, we're no longer trusting machines just to *do*
something, but to *decide* what to do and when to do it.
The next generation will grow up in an age where it's
normal to be surrounded by autonomous agents, with
or without cute names.

—RACHEL BOTSMAN

AI is how we can take all of this information and tell
you what you don't know about your health.

—JUN WANG

WHEN SIRI WAS INTRODUCED TO iPHONES IN 2011, IT WAS
great fodder for humor. Even a couple of years later, when I
visited the old *Colbert Report,* Stephen asked Siri, "Am I dying?" And
she responded, "I really can't say." There was a lot that Siri couldn't
do, and although 95 percent of iPhone owners tried Siri at some
point, for many would-be users it made a bad first impression and
they gave up.[1] Then, in 2015, Microsoft's Cortana arrived; soon
enough we were getting AI tips on our phone about traffic patterns

or that it was time to leave for the airport. In 2016 Google's virtual assistant was released, creatively called Google Assistant, with, as you might expect, the widest range of search commands. By the end of 2016, more than 40 percent of smartphone owners said they had used one of these assistants.[2] We're getting ever more used to using an artificially intelligent personal assistant.

I skipped over Amazon's Echo and Dot voice-controlled devices, which we know as Alexa, because they seemed to have taken the world (or at least the United States) by storm. Back in 2011, Jeff Bezos described his vision for the Alexa system: "A low-cost, ubiquitous computer with all its brains in the cloud that you could interact with over voice—you speak to it, it speaks to you."[3] Although introduced in late 2014 to its Prime members, it took a couple of years for Alexa's popularity to soar. By the end of 2016, however, Amazon couldn't keep up with production as Echo completely sold out, by then having established coresidence in more than 6 million US homes. Already 2016 was known by some aficionados as the year of "conversational commerce"—it had become ubiquitous enough that 250,000 people asked for Alexa's hand in marriage that year.[4] As of 2018, devices running Alexa account for more than 70 percent of all voice-powered AI devices, which are used by more than 60 million Americans.[5] They have qualified for tech unicorn status—the rare product that fundamentally changes the way we live.[6] The only other technology in US history that has had uptake by one in four Americans within two years was the iPhone in 2007 (Figure 12.1).

Why did this breakout of the voice platform for personal assistants occur when it did?[7] In retrospect, it should have been obvious that it would happen: humans engage far more naturally through speaking than they do by typing on a keyboard. As my friend Ben Greenberg at WebMD said, "It's pretty darn likely that our grandkids will laugh at us for ever using a keyboard."[8] But it goes well beyond that. Whether it's in English or Chinese, talking is more than two to three times faster than typing (both at the initial speech-transcription stage as

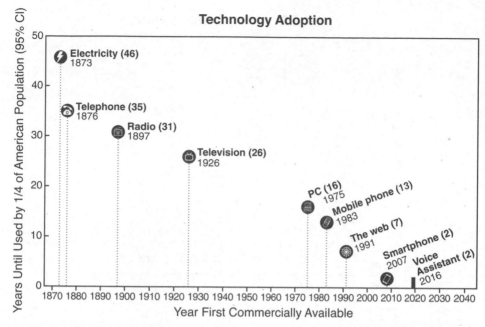

FIGURE 12.1: The time it takes from introduction of a new technology to adoption by one in four Americans. Source: Adapted from: "Happy Birthday World Wide Web," *Economist* (2014): www.economist.com/graphic-detail/2014/03/12/happy-birthday-world-wide-web.

well as when edited via keyboard) and has a significantly lower error rate in Chinese, a difficult language to type (Figure 12.2). It wasn't until 2016 that speech recognition by AI came into its own, when Microsoft's and Google's speech recognition technologies matched our skill at typing, achieving a 5 percent error rate. By now, AI has surpassed human performance.

Voice recognition has other advantages. There's no need for ID and passwords, avoiding the hassle of going through apps (that seem to need updating every day). All these factors make voice faster, hands-free, easy to use, and inexpensive. No wonder there's since been a major proliferation with other devices that integrate machine learning with natural-language processing including Google Home, Apple HomePod, Baidu's DuerOS, Clara Labs, x.ai, DigitalGenius, Howdy, Jibo, Samsung, LingLong DingDong (you wake it up by saying

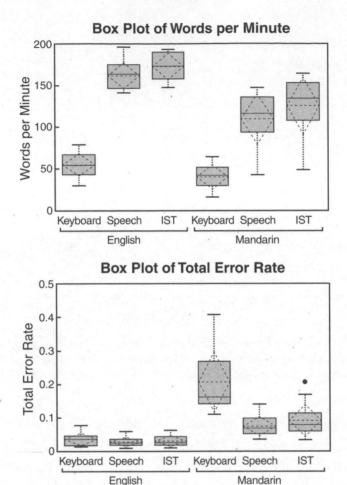

FIGURE 12.2: How fast we can talk versus type, in either English or Mandarin. The error rates are significantly lower for voice in Mandarin. Source: Adapted from: S. Ruan, *Speech Is 3x Faster Than Typing for English and Mandarin Text Entry on Mobile Devices,* arXiv (2016): http://hci.stanford.edu/research/speech/paper/speech_paper.pdf.

DingDongDingDong!) and many more to compete with Alexa. It is projected that 75 percent of American homes will have at least one Alexa or other voice personal assistant by 2020.[9] Many of my colleagues now have several, putting an Echo or Dot in different rooms of the house for convenience.

But what is that really providing now? *MIT Tech Review* touted that, "in the same way that smartphones have changed everything from dating etiquette to pedestrians' walking speed, voice-based AI is beginning to upend many aspects of home life."[10] But what it's upending is not anything particularly noteworthy—instead of peering at a smartphone screen, a user is talking to a cylinder. The tasks that these assistants accomplish—like shopping, playing music, turning down the lights, telling (corny) jokes, taking notes, predicting the weather, or ordering an Uber, toilet paper, or takeout, even faking a fart—are readily accomplished by other means. There's certainly no real, meaningful conversation to be had with a cylinder that has no understanding of the world. Microsoft's Xiaoice in China has had one of the longest conversations, but that's merely a machine metric, not an approximation of genuine human interaction. Indeed, fifty years after *2001* introduced HAL with a voice, we still haven't reached a convincing facsimile of extended human conversation. Amazon has more than 3,000 engineers working on Alexa. But when I asked Pedro Domingos, the University of Washington leading computer scientist, about it, he told me this brief story: "I went to this Amazon workshop where they gave me one, and I just set it up at home. The biggest fan is my kid. He likes to ask Alexa to tell him riddles. That's one of its skills. But then if you ask Alexa one of its own riddles, it doesn't even understand the question, let alone answer it."[11] Yet, when Google Duplex in 2018 demonstrated its humanoid chatty ability with words like "gotcha," "mmhmm," and "uhm" for making a restaurant reservation and other real-world tasks, this window dressing was viewed as a sign of great things to come.[12]

It's not that I don't think that these devices will get smarter and more conversant over time. After all, they are constantly learning from billions of voice interactions. Alexa has a two-year frontrunner edge and has cornered the market; with Amazon's AI expertise, there's no question that its voice assistant powers will be enhanced. The Alexa Prize, a $3.5 million competition (announced in 2018) to make Alexa chat like a human for twenty minutes without messing up, will likely help too.[13] There are also now tens of thousands of "skills" (the equivalent of a smartphone app) that Amazon and its open-source developers have added to Alexa's capabilities. And, even though it's not so easy to learn new cultures to provide voice support, Amazon has Alexa operational in German, Japanese, and French, and has other languages in the works.

There are some other particular advantages to the voice platform for AI. One clear advantage is for the blind and visually impaired, of whom there are more than 250 million people in the world, including, in the United States, almost 600,000 children under eighteen and more than 3 million individuals over age sixty-five.[14] Microsoft's Seeing AI is a free app that recognizes faces, scans supermarket food barcodes, identifies money, and reads handwriting. But Alexa takes this support further in the home environment for everyday tasks like dictating messages for texts and e-mails, finding TV shows to watch, or assessing one's outfits and appearance (via Echo Look). It is complemented by technologies like Aira Tech or MyEye, which are smart glasses with a camera, sensors, and network connectivity.

Let's not forget that there are about 780 million adults in the world who can't read or write, and there are remarkable translation capabilities now available that break down the barriers of interlanguage communication. For example, in China, more than 500 million people are using iFlytek's app, which turns Chinese speech into English text messages and vice versa.[15] There are even voice apps that have been developed for noisy environments, such as Xiaofeiyu for drivers in China. (Similar to voice recognition, deep neural networks are now mastering lip reading to help the deaf.)[16]

Nonetheless, there are clearly downsides that need to be underscored. Even though Alexa and the others have to be activated by the wake word—calling their name out—having a listening device in your home surely qualifies as creepy. It reminds me of Mark Zuckerberg putting tape over his laptop webcam, fearing that someone could be watching. You can shut down the "always-on" setting to alleviate some privacy concerns, but when the devices are activated, we know that the companies are taping some of the conversations to train and improve their platform, even though users have the ability to delete all the content (something that few people bother to do). That's why Alexa has been dubbed "Orwell's Big Brother."[17] If the voice part wasn't creepy enough, in 2017 Amazon introduced new Echo plus camera spin-offs like Spot with a display, Show with a touchscreen, and Echo Look, which uses a machine algorithm to tell you whether your outfit is aesthetically pleasing and stylish.[18] There's even AI science to support this capability.[19] I wonder what might be Alexa's suggestion for where you can go to improve your wardrobe.

The hacking vulnerability of voice assistants has been shown with such techniques as the so-called dolphin attack, which used ultrasonic frequencies, too high for the human ear to hear, to seize control of voice-activated gadgets.[20] There has even been a murder case in which Amazon was forced to provide the Echo recordings obtained when it wasn't activated but only listening, fulfilling the legal descriptor of a "ticking constitutional time bomb" with respect to the First Amendment.[21] Unknowingly, a couple in Portland, Oregon, had their conversation recorded and their audio files sent to their contacts.[22] These examples portend the problems of not establishing data protection and privacy.

Nicholas Carr, who is known for bringing out the bad side of technology, had this to say: "Even as they spy on us, the devices offer sanctuary from the unruliness of reality, with all its frictions and strains. They place us in a virtual world meticulously arranged to suit our bents and biases, a world that understands us and shapes itself to our desires. Amazon's decision to draw on classical mythology in naming its smart speaker was a masterstroke. Every Narcissus deserves an Echo."[23]

There has also been much concern expressed for the adverse impact of voice assistants on children, who are especially susceptible to attachment to these devices.[24] Just the headline, "When Your Kid Tries to Say 'Alexa' Before 'Mama,'" conveys the worry of a young child bonding with a seemingly all-knowing, all-hearing, disembodied cylinder voice.[25]

Despite these significant issues and trepidations, voice will likely emerge as the preferred platform for virtual assistants. The Alexa technology now is not right for portability, since a home cylinder speaker doesn't go with you wherever you go. Ideally, there must be seamlessness, such as an automatic handoff between smartphone and speaker or, more likely, an enhanced smartphone platform (with an avatar, voice, and text modes) that assumes this capability. Amazon is already bringing Alexa to smartwatches and headphone sets.[26] There's a further issue of limited output by voice when there's lots of text, such as the menu choices for a recommended restaurant; that is best seen on a screen. A solution awaits the optimal interface between different devices and the problem of ambient noise. With this background on virtual assistants for general use, we're now ready to move on to the progress so far in health and medicine. Much of the effort to bring AI into medicine to date has been directed toward enabling doctors and clinicians, rather than patients or healthy people. With smartphones carried by nearly 80 percent of the population and dedicated voice assistants heading in the same direction, both are suitable platforms for supporting the medical needs of consumers. Let's turn to the groups that are working toward the day when we might look in the fridge and ask, "Alexa, what should I eat?"

THE VIRTUAL MEDICAL ASSISTANT TODAY

While many AI apps have been developed to promote health or better management of a chronic condition, all are very narrow in their capabilities. In Chapter 4, for example, I presented the story of the AliveCor watchband. It uses deep learning to peg the relationship between a person's heart rate and physical activity, to prompt the user to record an electrocardiogram if his or her heart goes off track, and to look for evidence of atrial fibrillation. AliveCor's watch exemplifies the type of assistance developed to date. I won't try to review every possible example here, but I will cover enough to give you a flavor of where we stand in this nascent phase of the medical AI coach. There's an important common thread here: no randomized, controlled trials have been shown to improve outcomes. Instead, the products have largely relied on small retrospective or observational studies. It's a major hole in their stories that needs to get filled in. Nevertheless, the developments are worth examining.

Diabetes has been a popular target. Onduo, a company formed by Verily and Sanofi, is perhaps furthest along because it combines smartphone AI recognition of food and meals, continuous glucose sensor data, and physical activity (or what is really just steps taken) to provide coaching via texts. Wellpepper combines an Alexa-based strategy with a scale and a foot scanner. Patients with diabetes step on the scale and have their feet scanned; these images are processed via a machine learning classifier to detect diabetic foot ulcers. Voice prompts are used in conjunction to gather additional data as well as to provide education and management tips.[27] Virta is an expensive ($400 per month) smartphone app service designed to reverse type 2 diabetes with remote guidance of the individual's glucose measurements, diet, exercise, and medications via algorithms.[28] Other start-ups like Omada Health and Accolade use a hybrid of AI and human coaches for diabetes management. Notably, the companies, including Dexcom, Abbott, and Medtronic, that make continuous glucose sensors do not have deep learning algorithms capable of factoring in nutrition, physical activity, sleep, stress, the gut microbiome, and other data that might help people manage their condition. Instead they currently use "dumb" rules-based algorithms (not unlike those in the twelve-lead electrocardiogram) to warn people that their glucose is going up or down, based solely on prior values.

I discussed DayTwo, and its personalized machine learning nutrition algorithm driven by one's gut microbiome with the aim of optimizing glycemic response to food, in depth already. Veritas Genetics, the company that was the first to offer whole genome sequencing for less than $1,000, acquired an AI company with the aspiration of combining one's genomic data with individualized nutritional guidance. But their idea "Alexa, should I eat this last piece of pizza?"[29] is a long way off, with our very limited knowledge of nutrigenomics. There are a number of AI efforts for weight loss, such as Lark, that used a smartphone chatbot to achieve modest weight loss in a small cohort.[30] Similarly, Vida's AI app for weight loss, diabetes, and blood pressure management touts its personalized action plan that tracks self-reported stress, hunger, and energy levels. Human coaching has been shown to be effective for many of these conditions, with companies like Noom and Iora Health, so that may serve as a basis for continued AI efforts, or perhaps a hybrid approach will turn out to be the best strategy.

A narrow, disease-specific approach is being pursued by Tempus Labs for cancer. As discussed in Chapter 7 with cancer doctors, this company is aggregating comprehensive data from patients, including demographics, a genome sequence of the tumor and individual RNA sequence, immune response, medical imaging, liquid biopsy circulating tumor DNA sequencing, and organoids, along with treatment and outcomes. Not only is the company collaborating with most of the US National Cancer Institute centers, in late 2017 it gained access to more than 1 million patients' data from the American Society of Clinical Oncology's CancerLinQ. This complements the National Cancer Institute centers since CancerLinQ's database reflects community oncology practice from more than 2,000 oncologists from over a hundred practice groups throughout the country. With this unprecedented collation of data, Tempus, along with collaborators at Precision Health AI, is developing algorithms for improving cancer treatment outcomes.[31]

Second Opinion Health introduced Migraine Alert, a smartphone app, in 2017. Intermittent migraine sufferers are prompted to collect data of their potential triggers such as sleep, physical activity, stress, and weather. Its machine algorithm learns the person's pattern from fifteen episodes (that's a lot of headaches) to predict an imminent one with 85 percent accuracy, giving the person time to take preventive medications instead of treating the headache after the onset of pain.[32]

ResApp Health uses a smartphone microphone to listen to a person's breathing. The machine learning algorithm can purportedly diagnose several different lung conditions—acute or chronic asthma, pneumonia, and chronic obstructive lung disease—with high (~90 percent) accuracy.[33] AI has been used for matching patients with primary-care doctors by integrating extensive characteristics of both groups, achieving high predictive accuracy for trust measures.[34]

There are also many AI chatbots (some that work through Alexa and Google Home) and smartphone apps that perform varied functions like checking symptoms, promoting medication adherence, and answering health-related questions. These include Ada, Florence, Buoy, HealthTap, Your.MD, MedWhat, and Babylon Health. When Babylon Health had a public relations event in 2018 and posted a white paper on its website comparing diagnosis by a chatbot with that of seven doctors, claiming superiority of the former, it was harshly criticized for both methodology and overreach.[35] In a similar vein, journalists from *Quartz,* with assistance of doctors, evaluated sixty-five Alexa skills and the health information it provided, concluding, "Alexa is a terrible doctor."[36]

One other group of AI offerings is dedicated to seniors. Interestingly, care.coach takes the form of a speech-powered, puppy avatar to interact with and help monitor aged individuals.[37] Aifloo, a start-up in Sweden, developed a wristband that, when used in conjunction with AI to detect risk of fall, can alert caregivers.[38] Such technology, while never fully replacing human touch and care, may be useful adjunctively, especially given the profound mismatch between the proliferating population of people of advanced age and the major constraints on and expense of facilities to care for them.

Taken together, you can readily see how limited efforts are for AI virtual health coaching today. Overall, there's a very narrow focus, with thin, short, limited data collection, scant validation, and lack of far-reaching objectives.

BUILDING THE VIRTUAL MEDICAL ASSISTANT
OF THE FUTURE

Creating more powerful virtual medical assistants is both a techno-logical and a political challenge. In fact, both for reasons that we've covered, but that I want to return to here, and for some we haven't, the biggest issues might just be the political ones. This is impor-tant, not just because such assistants seem cool, but because they represent one of the most important boons of deep medicine—to empower not just physicians to be better at what they do, but to help all of us be as good as we can be at taking care of our own health. We cannot actualize the full potential of deep medicine unless we have something like a virtual medical assistant helping us out. No human, whether doctor or patient, will be able to process all the data. That's the unfulfilled promise of AI machines. When they are patient centered, it's the best hope we have to scale algo-rithmic functionality to the substantial proportion of people who might want, or benefit from having, their data served up to them. As Richard Horton, the editor in chief of the *Lancet,* who often expresses his skepticism about technology, wrote, "Replacing the doctor with an intelligent medical robot is a recurring theme in science fiction, but the idea of individualised medical advice from digital assistants, supported by self-surveillance smartphone data, no longer seems implausible." But we've got so many missing, crit-ical pieces right now for this be assembled.

The virtual medical assistant's value can only be as good as the input data. As Jonathan Chen and Steven Asch aptly wrote, "No amount of algorithmic finesse or computing power can squeeze out information that is not present."[39] First, all of a person's health-related data needs to be incorporated, ideally from the prenatal phase and throughout one's life, seamlessly and continuously updated. Until now in medicine, there has been a reductionist point of view.[40] That was evident with the Human Genome Project, whose premise was that understanding genomic variation would inform each individual's risk of diseases and treatments. This exemplified linear thinking, not appreciating the complex nature of health and diseases, and the multidimensional interactions with our microbiome, immune system, epigenome, social network, and environment (and more). Getting all of a person's data amalgamated is the critical first step. It needs to be regarded as a living resource that has to be nurtured, fed with all the new and relevant data, be they from a sensor, a life stress event, a change in career path, results of a gut microbiome test, birth of a child, and on and on. All that data has to be constantly assembled and analyzed, seamlessly, without being obtrusive to the individual. That means there should not be any manual logging on and off or active effort required. That's not so easy: for example, as I experienced firsthand, there's no method for capturing what food we ingest without manual entry via an app or website. When I did that for two weeks along with exercise and sleep (detailed in Chapter 11), my only solace was that it would only last for two weeks. Any AI coach that would be learning over a period much longer than days could not be predicated on users having to work to input their data.

Many creative passive solutions have been proposed or are actively being pursued. When I worked as an advisor to Google, the company, in collaboration with University of Rochester biomedical engineers, had designed a toilet seat that took a person's blood pressure while sitting on it. I'm not so sure that would be the most representative reading. But they also had other means, such as getting the person's vital signs from standing on a scale or looking in the bathroom mirror, as unobtrusive ways to get useful data. In parallel, many companies have been working on spectroscopic or colorimetric smartphone apps to scan food, besides Onduo's smartphone AI image detection. These might help, if they turned out to be accurate, but they would still involve effort and forethought on the part of the individual, which makes it less appealing. There's not much appetite for adding even more smartphone engagement during our mealtime.

Newer smartwatches are collecting more data than ever, such as the Fitbit's Ionic or Versa, which get continuous heart rate, sleep, and physical activity. The problem with some of these data, which theoretically would be valuable inputs for the AI coach, is their quality. As we discussed, movement during sleep is just a proxy for brain activity and isn't as good as having brain electrical activity monitored (via electroencephalogram), and we know all too well that while steps may be counted by digital trackers, they are only good for some activities, such as walking, and not for others, for example, bicycling or swimming. The main point here is that quality of the input data is essential, and settling for a lack of it would compromise the meaningful output of the AI assistant.

In Figure 12.3, I've depicted the complexity of the deep learning model for coaching an individual's health. You can see there's truly "big data" for an individual, making it both a formidable challenge and ideally suited for this form of AI. It would likely require hundreds of hidden layers of a neural network to get to our desired output—real-time, accurate, predictive, valuable guidance to promote health. Some AI experts may find this single model oversimplified and unrealistic. But it is precisely such end-to-end deep networks that we need, with complex network architecture that will very likely require combining other, and even yet undeveloped, learning AI tools (as in the AlphaGo machine triumph described in Chapter 4, which combined deep learning, reinforcement learning, and Monte Carlo tree search).

In many ways, we really don't know what composes the "holistic" view of each individual, and that informative, panoramic picture likely varies considerably from one person to another. For example, what particular sensors are needed to prevent or manage a condition? Transcriptomics or epigenomics, for example, are not going to be the same throughout the body; instead, they are unique to a particular cell type, and we can't access most of them. There are thousands of metabolites that can be assayed for a person, requiring mass spectrometry, and only at considerable expense. In a similar vein, if we were to try to characterize a person's immune system, the data would apply to only one moment in time, and the data would

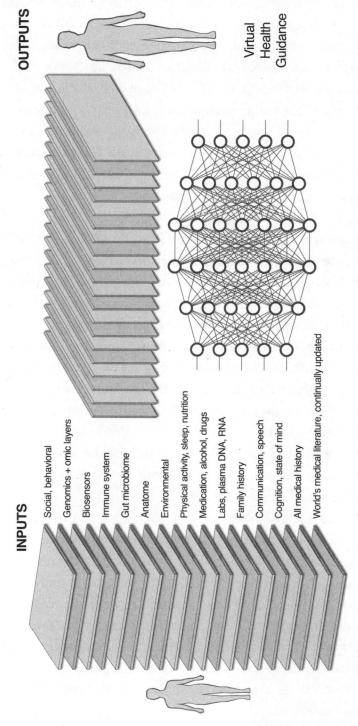

FIGURE 12.3: A schematic of a deep neural network with all of a person's data inputted, along with the medical literature, to provide the output of health coaching.

be complicated by the diverse approaches, such as autoantibodies, T and B cell repertoires, sequencing the histocompatibility complex, and flow cytometry, that are used to gather them. Which individuals should be monitored for plasma circulating tumor DNA for earliest detection of cancer, or for RNA signals for tracking of organ integrity (including brain, liver, kidney, and heart) and how often? What environmental data and sensors are appropriate to keep under surveillance, such as air quality or pollen count? I hope you can now appreciate the seemingly unlimited biologic, physiologic, anatomic, psychosocial, and environmental data to choose from.

The questions loom bigger than just these. The more we look at a person's metrics, the more likely we'll go down a rabbit hole and wind up with incidental findings, what my colleague Isaac Kohane nicknamed the incidentaloma. This is exemplified by whole body magnetic resonance or CT imaging, which invariably leads to seeing defects or abnormalities that require further evaluation, often requiring a biopsy, but proving to be benign cysts or nodules. This sets up a conundrum of collecting a deep stack of inputs for the AI coach, only to get an output riddled with false positive results, rather than achieve the intended goals of improved outcomes, prevention, or better management.

There are certainly myriad interactions in the human body of which we know very little. The network or system medicine approach would exploit AI to help us discover and understand how X connects with Y and Z, like a signal from the brain that influences blood pressure or from the gut microbiome that puts one at risk for cancer. Besides reductionist thinking in medicine that oversimplifies human health and diseases and that has such poor understanding of the "interactome," there's the daunting challenge of the fourth dimension—time. Each person is dynamic, constantly evolving, for better or worse, in some way, so whatever data are collated, we need to acknowledge that there are key constraints for interpretability. Setting the labels or ground truths for the neural networks could be exceedingly difficult.

Let's say we can use deep learning and other AI tools to work around the concern about false positives and to figure out what is the saturation point of data collection for any particular individual. And, further, let's say the challenges are successfully confronted with an AI medical coach that purports to promote health. The AI medical coach would have to be tested in a randomized, controlled trial to be accepted by the medical community and pass the ultimate clinical validation test. To date, only one company has embarked on this path. In China, iCarbonX, led by Jun Wang, who previously led China's largest genomic company (BGI), has attracted more than $600 million in funding and many corporate partners including SomaLogic, HealthTell, AOBiome, General Atomic Technologies Corp, Robustnique, Imagu, PatientsLikeMe, and two of China's largest insurance companies, AIA and China Life.[41] The data collection plan for iCarbonX is ambitious and partially overlaps with Figure 12.3: lifestyle, DNA sequencing, proteomics, metabolomics, immune system via autoantibodies, transcriptomics, the gut microbiome, continuous glucose monitoring, and the use of smart toilets and mirrors beyond smartphones to collect the data. The intent is to learn from a million people and develop the ultimate AI virtual medical assistant chatbot; the company's tag line is "Manage your life. Digitally." Some have speculated iCarbonX will need 10 million, not 1 million, people and far more capital than $600 million to execute this far-reaching mission. Nonetheless, it indicates that there is at least one major pursuit of the broader AI human health coach.

It may be too much to attempt, even for such a collaboration as the one iCarbonX has put together, to go after all of human health. Instead, it may make more sense to focus on particular acute conditions that might be prevented, like heart attack, asthma, seizures, or sepsis, or chronic conditions that would potentially be better managed, including high blood pressure, diabetes, depression, or different types of cancer. But this narrower approach would likely introduce human bias as to what data are useful as inputs, not taking advantage of the hypothesis-free discovery capabilities of the neural network. Nonetheless, it will probably be the default, a compromising way of moving forward in a non-holistic fashion. While this path for specific-condition coaching may accelerate success and validation, we shouldn't take our eye off the ball—the goal of overall health preservation.

Whether it is general or narrow, all the biomedical literature has to be continually ingested for the virtual medical coach to be maximally effective. When my orthopedist did not remember my congenital rare disease of osteochondritis dissecans or know that my postoperative physical therapy had to be different from the routine approach, a virtual assistant would only have been of help if it knew. Having all the medical literature at one's disposal is orders of magnitude more intricate than IBM Watson's ingestion of *Wikipedia*. Curating all the information and differentiating the quality from more than 2 million biomedical articles that are published each year does not, at least at this juncture, have an AI path to automation. AI extraction from text is a work in progress, certainly improving, and will be essential to support medical coaching.[42] An interim strategy might be to rely on a limited source of top-tier biomedical journals. My confidence that we'll eventually have the corpus of medical literature nailed by AI has been bolstered by conversations with Google's AI team and others.

There are major nonscientific challenges as well—the biggest is having all of a person's data. The notion that the electronic health record is the hallowed knowledge resource for each patient is a major problem. As we've seen, that couldn't be further from the truth. The EHR is a narrow, incomplete, error-laden view of an individual's health. This represents the quintessential bottleneck for the virtual medical assistant of the future. The input for deep learning about an individual relies on the completeness and veracity of the data, and the misconceptions about the value of American EHRs led me to write this tweet, which I titled "Your. Medical. Data." (Table 12.1).

I'll briefly review all twenty-four points. Because it's your body and you paid for the data, you ought to be its owner—not doctors

It's your body.

You paid for it.

It is worth more than any other type of data.

It's being widely sold, stolen, and hacked. And you don't know it.

It's full of mistakes that keep getting copied and pasted, and that you can't edit.

You are/will be generating more of it, but it's homeless.

Your medical privacy is precious.

The only way it can be made secure is to be decentralized.

It is legally owned by doctors and hospitals.

Hospitals won't or can't share your data ("information blocking").

Your doctor (>65 percent) won't give you a copy of your office notes.

You are far more apt to share your data than your doctor is.

You'd like to share it for medical research, but you can't get it.

You have seen many providers in your life; no health system/ insurer has all your data.

Essentially no one (in the United States) has all their medical data from birth throughout their life.

Your electronic health record was designed to maximize billing, not to help your health.

You are more engaged and have better outcomes when you have your data.

Doctors who have given full access to their patients' data make this their routine.

It requires comprehensive, continuous, seamless updating.

Access to or "control" of your data is not adequate.

~10 percent of medical scans are unnecessarily duplicated due to inaccessibility.

You can handle the truth.

You need to own your data; it should be a civil right.

It could save your life.

TABLE 12.1: My list of twenty-four reasons why you need to own your health and medical data.

and hospitals, which in all but one state in the United States legally own your data. If you owned and controlled it, you'd have a far better chance of preventing theft, hacking, and selling of your data without your knowledge. While many have pronounced the end of privacy, that simply will not work for medical data. The privacy and security of your data relies on its decentralization from massive server farms, the principal target of cyberthieves, to the smallest number possible— one person or a family unit being ideal—stored in a private cloud or blockchain platform. We've seen how every EHR has abundant mistakes that are perpetuated from one doctor visit to another, no less that we all have many different EHRs from all our different clinical encounters. And even if they were accurate, remember, too, that EHRs were designed for billing purposes, not to be a comprehensive resource of information about the individual.

The incompleteness is accentuated in the current era of sensors to track physiologic parameters, such as continuous glucose or heart rhythm, not to mention genomic data, which are absent as well. Of course, at this juncture, few people would want to have their genome data stored in a medical record that was owned by health systems or doctors, where they might fall into the hands of companies that sell life insurance or long-term disability plans. And we also have to acknowledge that the capture of other kinds of data—such as diet or even toilet-measured blood pressure—is not for everyone; many would have no part of it, and I certainly respect their point of view. Until we get medical data privacy protection and ownership straightened out, it's all the more reason not to trust comprehensive, continuous coalescence of your information.

In the current setup, getting hold of your data is remarkably difficult. Most American doctors don't willingly share their (really, your) office notes. Hospitals and health systems throughout the country are actively engaged in "information blocking," unwilling to share an individual's data for fear of losing control of the patient. One method to assure control is to use proprietary systems that don't produce interoperable files, as much a problem for any group seeking to make an assistant as it is for one health system to access the files of a rival one. Despite the US government's Health and Human Services Department calling out hospitals for this practice and the laws and regulations to prevent it, the problem persists.

I've argued, along with colleagues, that owning your medical data should be a civil right.[43] While I maintain that is the desired end goal and will likely be an inevitable reality someday in the United States, we can't wait decades for that to happen if we want to achieve the virtual medical coach's full potential. There are many countries around the world that have this configured appropriately. Take the little post-Soviet nation Estonia, profiled in the *New Yorker,* as "the digital republic:" "A tenet of the Estonian system, which uses a blockchain platform to protect data privacy and security, is that an individual owns all information recorded about him or her."[44] No one can even glance at a person's medical data without a call from the system overseers inquiring why it is necessary. The efficiencies of Estonia's health informatics system, in contrast to that of the United States, are striking, including an app for paramedics that provides information about patients before reaching their home and advanced telemedicine capabilities with real-time vital sign monitoring (with AI algorithms for interpretation) setting up doctoring from a distance and the avoidance of adverse drug-drug interactions. Although lacking as deep a digital infrastructure, other countries, such as Finland and Switzerland, provide medical data ownership rights to their citizens. Accordingly, there are existing models that prove this is not only feasible but also advantageous—the citizens of these countries have been outspoken in favor of having control and ownership of their health data. They are in a privileged, first-mover position for having laid the groundwork for the virtual medical assistant.

Then there is the form the assistant takes. Eventually, we will transcend talking to cylinders like the present-day Amazon and Google designs. In *The Power of Kindness,* my friend Brian Goldman, a Canadian emergency room doctor, devotes a chapter to the power of the "kindest" robots for communicating with people, particularly the aged with cognitive impairment.[45] The robots he observed in action with people in Japan, such as Telenoids from Osaka University and the Advanced Telecommunications Research Institute, are just the beginning. Hiroshi Ishiguro is the driving force there; he has gone on to create bots with stunning human facsimile, including their remarkable similarity to human hands.[46] Much like the leading character in the movie *Ex Machina,* Sophia, from Hanson Robotics in Hong Kong, is another example of a highly sophisticated, humanoid-appearing robot, and it's getting more interactive.[47] But the voice medical coach of the future needs to be eminently portable. That's why I think the human-like face avatars made by Soul Machines in Auckland, New Zealand, represent the prototype. These avatars have built-in AI sensors to detect a person's mood or fatigue, their eyes make close contact and track with you as you move, and their ability to engage in conversation is rapidly improving.[48] These avatars are already in use in kiosks by some airlines and banks; moving the software to a smartphone, tablet, or watch platform will be the next step. There are ongoing pilot studies in New Zealand for primary-care medical diagnosis and treatment.

Besides improved form factor, there are other issues to grapple with. To start with, there are plenty of people who would not want any part of an AI coach because of legitimate concerns about Big Brother and privacy, no matter how strong the assurance that their data would be secure. If indeed the AI coach improved outcomes and reduced costs, employers and health insurers would want to have these devices routinely used, creating tension and ethical concerns for many who desire and are entitled to autonomy. Even though it's really just software and algorithms, the cost of having an AI coach may be high and exacerbate the serious problems we already have with health inequities.

Much of the virtual medical assistant's ultimate success will be predicated on changing human behavior because so much of the burden of disease is related to poor lifestyle. As Mitesh Patel and colleagues asserted, "The final common pathway for the application of nearly every advance in medicine is human behavior."[49] There's no shortage of pessimism here. Take Ezekiel Emanuel, who wrote, "There is no reason to think that virtual medicine will succeed in inducing most patients to cooperate more with their own care, no matter how ingenious the latest gizmos. Many studies that have tried some high-tech intervention to improve patients' health have failed."[50] We've learned so much about behavioral science in recent years, but we still know relatively little about making people's lifestyle healthier. One of the leaders in this field is Theresa Marteau at Cambridge University, who points out that we stop swimming when we see a sign warning of shark-infested waters, but we don't respond to warnings for lifestyle improvement.[51] She and many other leaders contend that changing behavior requires targeting unconscious mental processes, the subtle, physical cues that shape our behavior—the nudges. We have yet to find the effective nudges that will durably change one's health habits, despite such tactics as financial or other incentives, gamification, or managed competitions. But we're getting better at using models to predict both online and offline behaviors, which may prove useful for identifying which people will likely be responsive or refractory.[52] To think that the pinnacle of the machine era for promoting health could be undone by human nature is sobering. But that potential obstacle needs to be confronted and hopefully superseded if the concept of a virtual medical coach will ever become a transformative, everyday reality. A new study from Finland of more than 7,000 individuals who were given their genetic risk score for heart disease is especially encouraging—at eighteen months follow-up, there was a striking proportion of participants who stopped smoking (17 percent) and lost weight (14 percent) among those with the highest risk.[53] The results contradict the notion that "personalizing risk information" will not prove effective.[54] Perhaps some combination of AI nudges, individualized data, and incentives will ultimately surmount this formidable challenge.

Today the self-driving car is viewed as the singular most advanced form of AI. I think a pinnacle of the future of healthcare will be building the virtual medical coach to promote self-driving healthy humans. Acknowledging there's no shortage of obstacles, I remain confident it will be built and fully clinically validated someday. If we, as humans, can put a man on the moon, develop the Internet, and create a Google map of Earth, there's no reason we won't be able to achieve this goal. Let me give some futuristic examples of what this might look like.

"Bob, I've noticed that your resting heart rate and blood pressure have been climbing over the past ten days. Could you pull out your smartphone retina imaging app and take a picture?"

"OK, Rachel, here it is."

"Bob, your retina doesn't show any sign of your blood pressure being out of control. So that's good. Have you been having any chest tightness?"

"No, Rachel."

"With your genomic risk profile for heart disease, I just want to be sure that isn't what is going on."

"Thanks, Rachel. I had some peculiar sensation in my jaw the last time I was on the treadmill, but it went away after a few minutes."

"Bob, that could be angina. I think an exercise stress test would help sort it out. I've looked at your schedule for next week and provisionally set this up with Dr. Jones on Tuesday afternoon at 4 P.M. on your way home from work if that's OK."

"Thanks, Rachel."

"Just remember to bring your running shoes and workout clothes. I'll remind you."

"David, I'm having some discomfort in my belly."

"I'm sorry to hear that, Karen. How long have you felt this sensation?"

"About two hours, and it seems to be getting worse, David."

"Karen, where are you feeling it?"

"It's on my right side, David."

"What and when did you eat last?"

"I had a hamburger, French fries, and iced tea at one."

"OK, Karen, do you have any other symptoms like nausea?"

"No, David, just the belly pain."

"OK, let's get out your smartphone ultrasound probe and place it on your belly."

"I've placed it, David."

"Karen, the images are not satisfactory. You need to move the probe up and more rightward."

"Is that it, David?"

"Yes, much better. I see that you have gallstones, which may be the explanation for the discomfort. Your family history of your mother goes along with it, as does your genomic risk score."

"That makes sense, David."

"Let me get Dr. Jones on the line to see what he recommends. The good news is that by ultrasound the stones look like they'll be dissolvable by medications."

"Randy, I just got your gut microbiome data to review."

"OK, Robin, what does it show?"

"The dominant bacteria is *Bacteroides stercoris*. It's twenty times more present in you than the general population. I just reviewed the literature, and there was a publication last week in *Nature*. It says this will cause spikes in your glucose after eating carbohydrates."

"Robin, I was worried because my glucose sensor has been showing many spikes after eating and I'm predisposed to become a diabetic. And you know I lost 10 pounds and have been exercising more in the past month."

"Right, Randy. Let me consult with the medical experts at YouBiome and see what microbiome modulation they recommend. Be back shortly."

[Your favorite music is played in the interim five minutes.]

"Randy, they say there's no need to take a PDQ probiotic prep. They advise a low carb diet for at least four weeks and then to reassess your glucose."

"OK, Robin."

"Randy, there was another bacterium that was present called *S. fecalis* that suggests you have an increased risk of colon cancer. Your last colonoscopy was seven years ago. I can make an appointment, or would you prefer taking a blood sample for tumor DNA?"

"I'll go with the blood sample. That damn prep for colonoscopy is torture."

"I've ordered the kit. It will arrive Wednesday."

"Sarah, how is your breathing?"

"It's fine, Katie."

"You are approaching a geo hot spot for asthma attacks, Sarah."

"Thanks for warning me."

"Sarah, let's do a lung function test."

"OK, Katie. . . . I just blew into the exhaler."

"Got that, Sarah. Your nitric oxide level is low, and your forced volume is low. Suggest taking two puffs of your inhaler."

"Done, Katie. And I see you have provided a detour from Front Street."

"It should only take two minutes longer."

"What do you suggest about my lung function?"

"Sarah, it looks like a combo of less exercise and high pollen count. The pollutant environmental sensor at home and work is stable, not trending up."

"I'll get walking more, Katie."

"John, I got your oxygen saturation last night, dipping to 67."

"Ann, I forgot to put on my BiPAP mask."

"John, your blood pressure went up to 195 at the time, and it was high all night long, mean of 155 systolic."

"So not just my sleep apnea, Ann?"

"No, John. Your weight gain of 12 pounds and no exercise in the past week is probably contributing."

"That's because my back was out of whack and I sat around eating."

"Yes, I have been warning you about that!"

"OK, Ann, That's enough. I quit. You're fired."

I hope these examples give a sense of where the field can go. I've tried to emphasize the need for holistic data and backup doctors and human experts. The virtual medical coach will ultimately prove to be a real boon for consumers, even though it's years away.

Now we're ready to turn to the final chapter in *Deep Medicine,* when we use the future to bring back the past.

chapter thirteen

DEEP EMPATHY

In learning to talk to his patients, the doctor may talk himself back into loving his work. He has little to lose and much to gain by letting the sick man into his heart.

—Anatole Broyard

By these means we may hope to achieve not indeed a brave new world, no sort of perfectionist Utopia, but the more modest and much more desirable objective—a genuinely human society.

—Aldous Huxley

I N THE FALL OF 1975, I ENTERED MEDICAL SCHOOL WITH NINETY other students, most of us fresh out of college. We were an idealistic lot. *Marcus Welby, M.D.,* about the kind family doctor with the ultimate bedside manner, was the hot medical TV show at the time, and *Dr. Kildare* reruns were still showing frequently. It was a simpler world of medicine—one with time to have a genuine relationship with patients. There were few medical procedures or fancy scans (beyond X-rays) or lab tests that could be ordered. Notes of a visit or hospital rounds were handwritten in the chart. A clinic appointment for a new patient was slotted for one hour minimum and return visits for thirty minutes. There was no such thing as a retail clinic. Or relative value units for a doctor's performance. Or monthly productivity reports for each doctor. There were few hospital or clinic administrators. There was no electronic health record, of course, and its requirement of spending twice as many hours with the computer as with patients. Even typewriters in medical facilities were not to be found. The term "health system" hadn't been coined yet. Throughout the United States, there were fewer than 4 million jobs in healthcare. We spent less than $800 per patient per year for healthcare; this accounted for less than 8 percent of the country's GDP.[1]

What a difference forty years make. Medicine is now a big business—the biggest business in the country. There are more than 16 million healthcare jobs in the United States (the leading source of employment in the country and in most cities, as well), and many "nonprofit" health systems have top-line revenues that tower into double-digit billions. We now spend more than $11,000 per person for healthcare and over $3.5 trillion per year, approaching 19 percent of the gross domestic product. Some medications and therapies cost more than $1 million per treatment, most new drugs for cancer start at more than $100,000 for a course, and many specialty drugs are about $2,000 per month. You can adjust the numbers for inflation and for the growth and aging of the population, and you'll quickly see that it's a runaway train. Health systems now have remarkable investment assets, like Kaiser Health with over $40 billion, Ascension Health with more than $17 billion, and Cleveland Clinic with over $9 billion.[2]

Along with the explosive economic growth of healthcare, the practice of medicine has been progressively dehumanized. Amazingly, ninety years ago, Francis Peabody predicted this would happen: "Hospitals . . . are apt to deteriorate into dehumanized machines."[3] (Parenthetically, if you read one paper cited in this chapter, this would be the one.) Rather than all the talk of "personalized" medicine, business interests have overtaken medical care. Clinicians are squeezed for maximal productivity and profits. We spend less and less time with patients, and that time is compromised without human-to-human bonding. The medical profession has long been mired in inefficiency, errors, waste, and suboptimal outcomes. In recent decades, it has lost its way from taking true care of patients. A new patient appointment averages twelve minutes, a return visit seven. Long gone are the days of Marcus Welby.

AI is going to profoundly change medicine. That doesn't necessarily mean for the better. The applications of the technology may be narrow and specialized today, with many of its benefits still in the promissory stage, but eventually it will affect how everyone in medicine—not just pattern doctors, like radiologists, pathologists, and dermatologists, but every type of doctor, nurse, physician assistant, pharmacist, physical therapist, palliative care provider, and paramedic—does their job. We will see a marked improvement in productivity and efficiency, not just for people but for operations throughout hospitals and clinics. It will take many years for all of this to be actualized, but ultimately it should be regarded as the most extensive transformation in the history of medicine. The super-streamlined workflow that lies before us, affecting every aspect of healthcare as we know it today in one way or another, could be used in two very different, opposing ways: to make things much better or far worse. We have to get out in front of it now to be sure this goes in the right direction.

THE GIFT OF TIME

One of the most important potential outgrowths of AI in medicine is the gift of time. More than half of all doctors have burnout, a staggering proportion (more than one in four in young physicians) suffer frank depression.[4] There are three hundred to four hundred physician suicides each year in the United States.[5] Burnout leads to medical errors, and medical errors in turn promote burnout. Something has to give. A better work-life balance—including more time with one-self, with family, friends, and even patients—may not be the fix. But it's certainly a start.

Time is essential to the quality of care patients receive and to their health outcomes. The National Bureau of Economic Research published a paper in 2018 by Elena Andreyeva and her colleagues at the University of Pennsylvania that studied the effect of the length of home health visits for patients who had been discharged from hospitals after treatment for acute conditions. Based on more than 60,000 visits by nurses, physical therapists, and other clinicians, they found that for every extra minute that a visit lasts, there was a reduction in risk of readmission of 8 percent.[6] For part-time providers, the decrease in hospital readmission was 16 percent per extra minute; for nurses in particular it was a 13 percent reduction per minute. Of all the factors that the researchers found could influence the risk of hospital readmission, time was the most important.

In 1895, William Osler wrote, "A case cannot be satisfactorily examined in less than half an hour. A sick man likes to have plenty of time spent over him, and he gets no satisfaction in a hurried ten or twelve minute examination."[7] That's true 120 years later. And it will always be true.

David Meltzer, an internist at the University of Chicago, has studied the relationship of time with doctors to key related factors like continuity of care, where the doctor who sees you at the clinic also sees you if you need care in a hospital. He reports that spending more time with patients reduced hospitalizations by 20 percent, saving millions of dollars as well as helping to avoid the risks of nosocomial infections and other hospital mishaps. That magnitude of benefit has subsequently been replicated by Kaiser Permanente and Vanderbilt University.[8]

These studies demonstrate the pivotal importance of the time a clinician spends with a patient. Not only does a longer visit enhance communication and build trust, it is linked with improved outcomes and can reduce subsequent costs. It's like an up-front investment that pays big dividends. That is completely counter to the productivity push in healthcare, where clinicians are squeezed to see more patients in less time. Of course, saving that money takes the doctor's time. One study, called the Healthy Work Place, of 168 clinicians in thirty-four clinics, demonstrated that pace of work was one of the most important determinants of job satisfaction.[9] A fascinating 2017 paper by psychologist Ashley Whillans and her colleagues, titled "Buying Time Promotes Happiness," showed that time saving resulted in greater life satisfaction. The people studied were diverse, drawn from representative populations of the United States, Canada, Denmark, and the Netherlands, as well as a separate group of more than eight hundred Dutch millionaires. The increased happiness derived from purchasing time was across the board, independent of income or socioeconomic status, defying the old adage that money can't buy happiness.[10] The ongoing Time Bank project at Stanford University's medical school shows how this works. The Time Bank is set up to reward doctors for their time spent on underappreciated work like mentoring, serving on committees, and covering for colleagues. In return, doctors get vouchers for time-saving services like housecleaning or meal delivery, leading to better job satisfaction, work-life balance, and retention rates.[11]

Like my classmates back in 1975, most people who have gone into the medical profession are motivated by, and feel privileged to have, the ability to care for patients. To a large degree, the rampant disenchantment is the result of not being able to execute our charge in a humanistic way. David Rosenthal and Abraham Verghese summed it up so well:

In short, the majority of what we define as "work" takes place away from the patient, in workrooms and on computers. Our attention is so frequently diverted from the lives, bodies, and souls of the people entrusted to our care that the doctor focused on the screen rather than the patient has become a cultural cliché. As technology has allowed us to care for patients at a distance from the bedside and the nursing staff, we've distanced ourselves from the person-hood, the embodied identity, of patients, as well as from our col-leagues, to do our work on the computer.[12]

AI can help achieve the gift of time with patients. In 2018, the Institute for Public Policy Research published an extensive report on the impact of AI and technology titled "Better Health and Care for All," projecting the potential time freed up for care of patients will average more than 25 percent across various types of clinicians.[13] One of the most important effects will come from unshackling clini-cians from electronic health records. At the University of Colorado, taking the computer out of the exam room and supporting doctors with human medical assistants led to a striking reduction in physi-cian burnout, from 53 percent to 13 percent.[14] There's no reason to think that using natural-language processing during patient encoun-ters couldn't have the same effect. But the tech solution per se won't work unless there is recognition that medicine is not an assembly line. As Ronald Epstein and Michael Privitera wrote in the *Lancet,* "Physicians, disillusioned by the productivity orientation of admin-istrators and absence of affirmation for the values and relationships that sustain their sense of purpose, need enlightened leaders who recognize that medicine is a human endeavor and not an assembly line."[15] They've got it mostly right: we need everyone on board, not just leaders. If the heightened efficiency is just used by administrators as a means to rev up productivity, so doctors see more patients, read more scans or slides, and maximize throughput, there will be no gift of time. It's entirely possible that this will happen: it was, after all, doctors themselves who allowed the invasion of grossly inadequate electronic health records into the clinic, never standing up to compa-nies like Epic, which has, in its contracts with hospitals and doctors, a gag clause that prohibits them from disparaging electronic health records or even publishing EHR screenshots.[16] This time it will be vital for doctors to take on the role of activists.

Unfortunately, doctors' activism is unlikely to be supported by professional medical organizations, at least not in the United States. For one thing, there is no singular representation for doctors: the American Medical Association membership is not even one-third of practicing physicians.[17] Worse, even that representation is hardly real: professional medical groups function predominantly as trade guilds to protect reimbursement for their constituents. There is a lot of capital available for potential influence, however. Of the top seven US government lobbyists in 2017, four were healthcare entities: Pharma Research and Manufacturers ($25.8 million), Blue Cross Blue Shield ($24.3 million), American Hospital Association ($22.1 million), and the American Medical Association ($21.5 million).[18] These days, unfortunately, it's used to protect financial interests, not the interests of patients or clinicians.

But even as tech gives doctors more time, that will not be enough. It is, however, the root of several necessary changes to how physicians are able to think about, and interact with, their patients that must be achieved if medicine will ever truly be deep.

BEING HUMAN

There's a major shortage of empathy in medicine today, only part of which is related to inadequate time.

Ironically, Matthew Castle, a physician in England, published "Burnout," an essay in which he projected himself as an AI machine doctor in 2100. He had plenty of deep learning intelligence, complete molecular and neuropsychiatric profiling of each patient, command of the entire biomedical literature, and the ability to perform thousands of simultaneous consultations. With all that data and AI, you'd think everything would be utopian, but the company he works for is demanding humanistic qualities. He winds up with burnout and requests a six-month sabbatical because "the problem is your requirement to develop empathy"! He writes, "It doesn't matter how powerful the human or machine software: ask it to do something impossible, and it will fail."[19]

As machines get smarter, humans will need to evolve along a different path from machines and become more humane. In Figure 13.1, I've tried to portray this point. Human performance is unlikely to change materially over time. But machines will progressively outperform humans for various narrow tasks. To take humans to the next level, we need to up our humanist qualities, that which will always differentiate us from machines. Notably, human empathy is not something machines can truly simulate, despite ongoing efforts to design sociable robots or apps that promote empathy. Yes, it's true that AI that seeks to detect human emotions like anger, sadness, fatigue, and distraction is being pursued.[20] Some capability for empathy is being built into virtual humans manufactured by the most advanced robotics companies, but even their AI experts admit there will always be a gap, the inability to "imbue such a machine with humanness"—that ineffable presence the Japanese call *sonzai-kan*.[21] And empathy is just one of the essential features of being human. Let me add the ability to love, laugh, cry, dream, be afraid, grieve, feel joy, trust in and care for one another, suffer, explore, tell stories, inspire, be curious, creative, grateful, optimistic, kind, express emotions, understand, be generous, and respectful. And to be adaptable, innovative, intuitive, have common sense, culture, and the ability to abstract and contextualize. To have a soul. And much more.

Brian Christian, an AI expert, wrote about being human in *The Most Human Human:* "To be human is to be a human, a specific person with a life history and idiosyncrasy and point of view; artificial intelligence suggests that the line between intelligent machines and people blurs most when a puree is made of that identity." Which is all the more reason we cannot allow this blurring to occur.

An important clause in the Hippocratic Oath holds that "sympathy and understanding may outweigh the surgeon's knife or the chemist's drug." Empathy is the backbone of the relationship with patients. In a systematic review of 964 original studies examining the

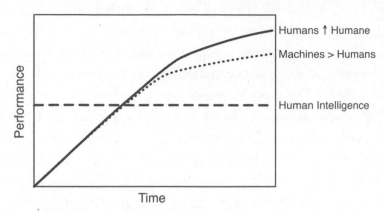

FIGURE 13.1: Over time, human intelligence and performance is not likely to change or improve in any significant way. Machines, for many narrow tasks, will increasingly achieve superhuman performance. Our charge ahead in medicine is to take our humanistic qualities to the next level.

effects of doctors' ability to empathize, there was an unequivocally positive relationship between empathy and improvements in clinical outcomes, patient satisfaction, adherence to recommendations and prescriptions, and reduction of anxiety and stress.[22]

Empathy is crucial to our ability to witness others who are suffering.[23] Ironically, as doctors, we are trained to avoid the s-word because it isn't actionable. The *American Medical Association Manual of Style* says that we should "avoid describing persons as victims or with other emotional terms that suggest helplessness (afflicted with, suffering from, stricken with, maimed)." Thomas Lee, writing in the *New England Journal of Medicine,* argues that, although "human suffering can be considered (and preferably averted) in the abstract," it's important that "patients must generally simply 'have' a disease or complications or side effects rather than 'suffer' or 'suffer from' them." He claims that physicians ought to seek to "avoid the word 'suffering' even though we know it is real for our patients because the idea of taking responsibility for it overwhelms us as individuals—and we are already overwhelmed by our other duties and obligations." Perhaps it's not surprising, then, that there are billing codes, reimbursement rates, and pills for treating anxiety but none for alleviating suffering.

There's also no machine that will do it; alleviating suffering relies on human-to-human bonding; it requires time, and its basis is trust.

I recently saw a patient, a young woman who came for a second opinion because she had experienced multiple episodes of "aborted sudden death." The term alone conveys a coldness even worse than labeling people as having "heart failure"—a coldness that needs to be replaced with something warmer. The way we speak about our patients' suffering becomes the important words that affected people have to live with and think about every day of their lives. To prevent another episode of the life-threatening heart rhythm, my patient had a defibrillator implanted. This required an extensive amount of hardware to be inserted into the heart and body of this young lady. The procedure, in itself, had to be traumatic. But that was not the extent of her suffering. As she told me her fears and concerns, she said she and her husband wanted to have a baby, but. . . . She started sobbing, and continued to, as she struggled to get the words out, that she didn't want to pass along the bad gene to the baby. I started crying, too, fully identifying with her worry. Maybe, in part, because my daughter was pregnant at the time.

She was suffering, not just for what she went through, but for envisioning putting her daughter or son through the same. I held her hand and tried to comfort her. That was also to comfort me. After a couple of minutes, I told her we would sequence her genome to look for the mutation that might have caused the disordered heart rhythm. And if we found it, that could help in selecting an embryo to avoid passing it along to her children. Months later we were able to find the mutation; she and her husband were relieved to know that they may be able to conceive without this horrifying emotional anxiety and overhang. The experience made me think that the term "deep medicine" was fitting.

Given the importance of empathy for improving physician performance and psychosocial outcomes, it's vital to know whether empathy can be nurtured—or destroyed. Zak Kelm and his colleagues undertook a major analysis of sixty-four studies, ten of which they considered to be rigorously designed; in general, the studies showed that empathy can be cultivated among physicians.[24] Unfortunately, empathy scores decline during medical training, as empathy is challenged by the practice environment. As David Scales, a medical resident, noted, practitioners lack the time to care for patients as the doctors hope and patients deserve, with physicians blaming "the time pressure created by a billing system that promotes quantity of patients seen over quality, the lack of control over the chaotic work environment, and endless time spent on administrative tasks."[25] We also know that medical professionals generally have low scores on empathy quotient (EQ) tests. Altruists have EQs in the 60–70 range, artists and musicians in the 50s, doctors in the 40s, and psychopaths less than 10.[26] We even have seen the neuroanatomy of empathy defined and can pinpoint the precise part of the brain and circuits responsible, along with the biological, psychological, and social activation and suppression. What is encouraging, however, is that there is plasticity of the brain for the critical soft skills of empathy, compassion, and taking the perspective of another person, akin to the way it is possible to have hypertrophy of parts of the brain responsible for navigation, as has been found in London taxi drivers. For example, more than three hundred healthy adults (not doctors) underwent training meant to enhance presence (defined as attention and interoceptive awareness), affect (care, compassion, prosocial motivation, dealing with difficult emotions), and perspective (metacognition, perspective taking on one's self and others). Serial MRI imaging during the training revealed significant changes over nine months in brain morphology associated with each behavioral module.[27] So, there is hope—indeed, anatomical and empirical evidence—that empathy and soft skills can be fostered and that we could take on intensive initiatives to promote empathy in all clinicians. The healers, after all, need healing, too. It shouldn't take the escalating incidence of depression and suicide to move on this potential.

PRESENCE

Empathy is a start. But the problem with the patient-doctor relationship is much bigger than having or missing empathy. For the human-to-human connection to be deep and genuine, many factors are essential. A principal reason I asked my friend Abraham Verghese to write the foreword for this book is because he has been the pioneer for presence—the art and science of human connection—and started a major initiative to champion the field.[28] As Verghese puts it, "Being present is essential to the well-being of both patients and caregivers, and it is fundamental to establishing trust in all human interactions." He gave us his definitive definition: "It is a one-word rallying cry for patients and physicians, the common ground we share, the one thing we should not compromise, that starting place to begin reform, the single word to put on the placard as we rally for the cause. Presence. Period."[29]

Sharon Roman, a patient with multiple sclerosis, wrote that, "when the hands have become calloused and rough, the ears unhearing, and examinations start to feel like interrogations, it is time to reconsider" one's choice of doctor.[30] There isn't any question that our patients want doctors to be present, with intentional listening and undivided attention. That rarely occurs now. Rather than listening, doctors interrupt. Indeed, it only takes an average of eighteen seconds from the start of an encounter before doctors interrupt their patients. Eighteen seconds.[31] This desire to cut to the chase instead of giving the patient a chance to tell her narrative certainly matches up with the extreme time pressure that doctors and clinicians are facing. What a blown opportunity to get to know people, to observe their emotions in relating their concerns and symptoms and their own theories for what's wrong. It was the father of modern medicine, William Osler, who said, "Just listen to your patient; he is telling you the diagnosis." Likewise, my friend Jerome Groopman wrote a whole book—*How Doctors Think*—on the adverse impact of not listening, not giving patients a voice. When the journalist Andrea Mitchell reflected on her career, she said the advice that Tim Russert gave her was instrumental: "Always listen in between the seams of someone's answer," which equally applies in medicine.[32] We need patients to be free to be storytellers because, even as AI manages to synthesize notes and labs and imaging into something actionable, it will never be able to tell a patient's story the way a patient would. As doctors we are trained to take the history. But that's clearly the wrong concept; it preempts conversation, which is both giving and taking.[33] That's how the deepest and intimate feelings come out, and if there's one thing doctors wish, it's that "they had time to talk with their patients, knowing the value of such contact."[34]

An essay by a medical student upon seeing her first patient comes to mind. Julia Schoen wrote about her encounter with Mr. B, who was described by her team as "a 63-year-old man with HF-pEF and err . . . PHTN and COPD presenting with an acute CHF exacerbation." But Schoen was imagining Mr. B, who had heart failure, puffing and panting as he wheeled himself across the street. "I could hear his wet wheeze as he rested on the other side of the street," she wrote. "I wondered how many silently went out of their way to avoid him." She wondered about what patients want in their ideal healer; as she listened to him telling jokes and stories she felt she was the patient appreciating the beauty of life. This first patient encounter taught her about the primacy of "listening, learning and loving my patients."[35]

Schoen's story exemplified letting down her guard, tearing down the wall between two people, ultimately leading to a deep patient relationship. There are many ways that narrowing the gap can be accomplished. A new trend in some medical centers is for doctors to give their patients a card with their picture and details about their family, where they live, and their hobbies and nonmedical interests.[36] While this lies in direct opposition to how doctors have historically been trained, it represents the right future path of humanistic medicine.

Back in 1999, a few years after losing his son from leukemia, John Iglehart, the editor of the journal *Health Affairs,* wrote a brief introductory piece for a new series in the journal called "Narrative Matters" because "the voices of patients, their families, and their caregivers have often gotten lost in the relentless shuffle. . . . of big business."[37] Since then there have been hundreds of essays under that heading in that journal, along with a similar series in both the *Lancet* and the *Annals of Internal Medicine* (where it's called "On Being a Doctor"). I turn to them every week to nurture my own capacity for presence and empathy in the clinic.[38] One of my recent favorites was titled "You Don't Know Me." In it, a hospitalized man with a brain tumor keeps telling his doctor, Kate Rowland, that she doesn't know him. Sick and dying, the patient said, "This is not who I am." When she read his obituary, she remembered that she had his business card, and for ten years she has carried that card in the same coat pocket where she carried her own, as a constant reminder that the patient was right, that she really didn't know him.[39] And she's spot on: we hardly ever (maybe never) *really* get to know patients, but if there's little time, little presence, and little voice of the patient, there's no chance. I can assure you that there will be no AI that will truly *know* a person, which is hard and requires a dedicated human-to-human connection. AI will give us the time. But we need to make it happen.

Rita Charon, a physician at Columbia University who has pioneered narrative medicine, described how she changed her practice to achieve presence:

> I used to ask new patients a million questions about their health, their symptoms, their diet and exercise, their previous illnesses or surgeries. I don't do that anymore. I find it more useful to offer my presence to patients and invite them to tell me what they think I should know about their situation. . . . I sit there in front of the patient, sitting on my hands so as not to write during the patient's account, the better to grant attention to the story, probably with my mouth open in amazement at the unerring privilege of hearing another put into words—seamlessly, freely, in whatever form is chosen—what I need to know about him or her.[40]

Part of presence, too, is the power of careful, detailed observation. I was struck two decades ago when Yale's medical school announced a required course for students to learn the art of observation by spending time in an art museum.[41] Verghese gets that, too, writing in a "Narrative Matters" piece that "my tool is the medical gaze, the desire to look for pathology and connection, and it would seem there was no opportunity for that within a pigmented square of uniform color or a rectangle of haphazard paint splashes. But in me a profound and inward sort of observations was taking form." Abraham takes medical students to the art museum at Stanford to foster observational skills.[42]

These aren't just fanciful claims that Verghese and Charon are making. In 2017, a small group of first-year medical students at the University of Pennsylvania participated in a randomized trial of art training at the Philadelphia Museum of Art compared to controls without such exposure. The training consisted of six 90-minute sessions over a three-month period, and the outcome of observational skills, describing both art and medical images, showed a remarkable benefit.[43] David Epstein and Malcolm Gladwell wrote an editorial to accompany the paper, which they called "The Temin Effect," after the Nobel laureate Howard Temin, who not only discovered reverse transcriptase but also read deeply about philosophy and literature.[44] Their conclusion: "Taking would-be physicians out of the hospital and into a museum—taking them out of their own world and into a different one—made them better physicians."

Sarah Parker, a neurologist, wrote about an extraordinary example of human connection, empathy, and keen observation, without a single word uttered, in the face of tragedy:

The doctor walked out of one of his clinic rooms and told his nurse he thought he was having a stroke. By the time I saw him, he was nonverbal, globally aphasic, unable to move his right side, his brain filling with rapidly expanding hemorrhage. He didn't understand what I was asking him to do. He couldn't tell me what he was feeling, but he recognized my white coat. He recognized the tone in my voice. He recognized the expression on my face. He took my hand in his left hand and repeatedly squeezed it and looked me right in the eyes. There was a moment of connection. A moment where two people know what the other is thinking and feeling without a word passing between them. He knew this was bad. He knew I knew this was bad. He knew I was trying to help, but he knew that there wasn't much I could do. He was scared but also strong and courageous. He knew the situation and knew the likely outcome, and he was telling me that it was OK if it ended that way. That he knew I cared. It was a moment of peace. A man facing death, both afraid and aware. A man who was looking for that human connection. A man who had cared for others and comforted others his whole life was trying to comfort me while I tried to care for and comfort him.[45]

It's a bit ironic that the hero in the current hit medical show *The Good Doctor* is a surgical resident with autism who has a savant syndrome. Within seconds of looking at a patient's scan he makes the diagnosis and sees details that aren't apparent to other doctors.[46] We don't need to be savants to be far better observers. It takes time and can be bolstered by training. Besides, visiting art museums is enriching.

THE PHYSICAL EXAM

Observation doesn't just end with listening to a patient. It extends to the physical exam, which, in the essential laying on of hands, epitomizes the concrete sense of human touch and the intimacy of one human disrobing to be examined by another. The physicality of it may be the antithesis of an algorithm. Over the years I certainly have observed a ritual lost as respect for and execution of the physical exam degrades. Physicians are progressively getting out of touch—literally—with their patients. All too often my colleagues put in the record "WNL," which is supposed to mean "within normal limits," but, in reality, often can be described as "we never looked." It's too easy to order an echocardiogram or an ultrasound rather than taking the time to do the physical exam, in part knowing the inside look will provide more useful information. Similarly, it's almost routine these days not to ask patients to disrobe, even though it should be done to complete a proper exam. Using a stethoscope to listen with clothes between the patient and doctor is lame. The exam is crucial to earning a patient's trust. The outside look and feel is complementary to the inside look, if obtained. It is the essence of the human touch in medicine that cannot and should not be abandoned. As Verghese writes, "I find that patients from almost any culture have deep expectations of a ritual when a doctor sees them, and they are quick to perceive when he or she gives those procedures short shrift by, say, placing the stethoscope on top of the gown instead of the skin, doing a cursory prod of the belly and wrapping up in thirty seconds. Rituals are about transformation, the crossing of a threshold, and in the case of the bedside exam, the transformation is the cementing of the doctor-patient relationship, a way of saying: 'I will see you through this illness. I will be with you through thick and thin.' It is paramount that doctors not forget the importance of this ritual."[47] It's no surprise that Verghese has felt the strongest connection with physical and massage therapists, the only ones who really examined his body.[48]

Again, the lack of time is a dominant part of the explanation for these deficiencies. I wholeheartedly agree with Verghese's observation: "For the past two decades I've felt that in the United States we touch our patients less and less: the physical exam, the skilled bedside examination, has diminished to where it is pure farce."[49]

This makes me think of a patient of mine who came to see me after moving from Cleveland to San Diego. He had prior bypass surgery and an abnormal stress test with symptoms that could be construed as angina. Because he was in a rush to get this assessed, I saw him at the same time as another cardiologist colleague, who was going to perform the cardiac catheterization. By the time I got to the exam room, that doctor had already completed the physical exam, so he reviewed it with me in front of the patient and his wife. The four of us discussed the situation, agreed on the plan, and soon thereafter the patient went into the cath lab. He had a narrowing in a bypass graft that was successfully stented, and he was discharged the following morning. When I went to see him, I was expecting to see the happy gentleman whom I had known for many years. To my chagrin, however, I saw a distraught man who was upset with me. When I asked him why, he looked at me wryly and said, "You didn't examine me." I apologized; we made amends. But to this day that incident has stuck with me—it drove home how important the exam is to reassure and comfort a patient, even when someone else may have already done it or when, as a physician, I don't expect to get much information from the endeavor. And of course I was disappointed after my total knee replacement when, in the midst of my difficult recovery, my orthopedist didn't even examine my knee.

Recently, Michael Aminoff, a neurologist at UCSF, pondered the future of the neurologic exam:

> The neurologic examination requires time, patience, effort, and expertise and may have to be performed in difficult or unpleasant circumstances, whereas an imaging or laboratory study simply requires completion of a request form and the responsibility is passed to a colleague. Why, then, examine the patient? . . . An especially important aspect of the physical examination is that it establishes a bond between physician and patient, helping to establish a special relationship—of mutual understanding and respect—that is difficult to appreciate for those who themselves have never (yet) been a patient. The art of clinical neurology involves the ability to interact with patients on a human level and to relate any findings, whether clinical or investigative, to the context in which they were obtained. If medicine becomes dehumanized by technology, the quality of health care—or certain aspects of it—inevitably will deteriorate, in much the same way that voicemail, which facilitates communica-

tion, commonly engenders uncertainty, frustration, and impatience because direct human contact is lacking. The neurologic examination restores the physician-patient relationship and allows clinical problems to be viewed in context so that clinical common sense—imperiled by simplistic algorithmic approaches—can be used to manage them.[50]

To restore the primacy of the exam, we need to think of how technology has changed and adapt. During my residency at UCSF, I was trained by my medical hero, Dr. Kanu Chatterjee. We spent most of our time together (along with other trainees) at patient bedsides in the coronary care unit. We'd go in the room to talk with the patient and then take time to observe the chest and neck—to see whether the neck veins were elevated or the carotid artery pulse irregular, or whether the heart impulse was showing some sign of abnormal motion as seen from the chest wall—before doing any other part of the exam. After inspection, it was time for palpation, feeling the wrist artery pulse, the carotid artery pulse upstroke in the neck, and the heart's impulse in the chest. Then we would spend several minutes listening carefully to the heart sounds, especially for the splitting of the second heat sound (the dub in lub-dub). Listening in many positions, with the patient lying nearly flat, on his or her side, or sitting up (typically by maneuvering the bed), for a murmur, a rub, or a click. After this very slow, deliberate, systematic approach to the exam, Chatterjee typically was able to predict the pressure metrics for different chambers of the heart within 1 or 2 millimeters. I was totally sold on the importance of the exam from just that remarkable potential, but the patients found the in-depth scrutiny crucial as well. So, for the next few decades, whenever serving as a teaching attending in the hospital or with my fellows in clinic, I tried to emulate this meticulous exam. To be like Kanu, even though I could never approach his masterful capabilities.

Now that the stethoscope is 210 years old, and although it endures as the icon of medicine, it's time to reexamine the tools of the exam. The stethoscope is simply rubber tubing that doesn't record anything and, at best, can only transiently serve as a conduit for hearing body sounds. I could never share heart sounds with a patient in any meaningful way, even when there was a rare murmur that sounded like a washing machine. Patients couldn't know what the sounds represented. Now that there are smartphone ultrasounds, we can get a direct visualization of the heart rather than having to extrapolate it from sounds; we can capture and store the data as well as immediately share and interpret it with patients, enabling them to see and get a sense of what they look like inside. Like AI, this is technology that improves certain aspects of medicine and can directly enhance communication and promote bonding between the patient and doctor.

THE PATIENT-DOCTOR RELATIONSHIP

The fundamentals—empathy, presence, listening, communication, the laying of hands, and the physical exam—are the building blocks for a cherished relationship between patient and doctor. These features are the seeds for trust, for providing comfort and promoting a sense of healing. They are the building blocks that enable genuine caring for the patient and a doctor's professional fulfillment that comes from improving a person's life. All these humanistic interactions are difficult to quantify or digitize, which further highlights why doctors are irreplaceable by machines.

Patients seeking care are intrinsically vulnerable, setting up a conundrum when a patient and doctor first meet—to be open requires trust, but there's not much reason to trust someone you don't know.[51] At this moment of vulnerability and need for trust, patients confront doctors who are trained in medical school to keep an emotional distance from their patients. That's simply wrong. Without trust, why would people reveal their most intimate and sensitive concerns to a doctor? Or agree to undergo a major procedure or surgery, putting their lives in the doctor's hands?

An essential aspect of the relationship is the ability to deliver bad news. We should never delegate this to an algorithm. There is a mnemonic used to guide doctors for delivering bad news called SPIKES: s̲etting, a quiet private room; p̲atient perspective, understanding of the situation; i̲nformation, how much the patients and family want to know; k̲nowledge, letting the patient speak first after delivering the news; e̲mpathize with things like "I can't imagine how difficult this news must be for you"; and s̲trategize next steps. Of all the different fields in medicine, cancer doctors may have the highest burden for breaking bad news, as many as 20,000 times over a career.[52]

Danielle Ofri, a physician-author who is fully in touch with the essence of the patient-doctor relationship, wrote, "We are given unwitting front-row seats to the fortitude of humanity," and yet the medical profession encourages distance.[53] Thinking about each letter of a mnemonic won't be the way to exude human compassion. And as Ofri notes, when a patient dies, we doctors tell others that we "lost a patient," as if medicine were a search for a missing item in the lost and found. How many doctors attend a patient's funeral, even though little else could be a more exemplary way of communicating empathy? Gregory Kane imagined a future in which "archaeologists may survey the remains of our society and marvel at medical technologies evidenced by the collection of joint replacements, cardiac stents, valvular implants, and titanium plates among the remains in our places of burial." This might seem like the core of the legacy of modern medicine, but, having listened to a patient crying about lack of any contact from the doctor who had treated her husband for lung cancer, Kane had different thoughts on this matter: "It would be my hope that they would also identify in the written archives a condolence letter to note the personal connections that bound the physicians of the age to their patients and the surviving loved ones, providing evidence that we are truly human."[54]

The simple matter of reaching out to a family with a condolence letter can help reduce suffering, honor the person's life, and highlight human dignity. The converse can be illuminating and inspiring, too. It was instructive to read an essay called "The Greatest Gift: How a Patient's Death Taught Me to Be a Physician," by Lawrence Kaplan, a physician at Temple University. A patient's son sent Kaplan a note: "Thanks for everything you did for my father. It meant more than you can ever imagine." Accompanying the note was a picture of two saplings planted side by side, one in honor of the doctor and one in honor of the father. Kaplan described how the gift inspired him to relearn how to care for patients, and, to this day, the photograph sits prominently in his office as a reminder of what really matters.[55]

Fortunately, most doctor-patient interactions are not centered on the death of an individual, or even on the curing of a patient, but instead on healing. Verghese describes this dichotomy eloquently:

> We are perhaps in search of something more than a cure—call it healing. If you were robbed one day, and if by the next day the robber was caught and all your goods returned to you, you would only feel partly restored; you would be "cured" but not "healed"; your sense of psychic violation would remain. Similarly, with illness, a cure is good, but we want the healing as well, we want the magic that good physicians provide with their personality, their empathy and their reassurance. Perhaps these were qualities that existed in abundance in the prepenicillin days when there was little else to do. But in these days of gene therapy, increasing specialization, managed care and major time constraints, there is a tendency to focus on the illness, the cure, the magic of saving a life.[56]

Nearly a century ago Peabody wrote about this: "The significance of the intimate personal relationship between physician and patient cannot be too strongly emphasized, for in an extraordinarily large number of cases both diagnosis and treatment are directly dependent on it."[57] When there is a genuine, deep patient-physician relationship, healing comes easily and naturally. Patients will then believe doctors who say they will be there to support them, whatever it takes. It is what most patients crave, but is so hard to find in this day and age. That has to change; we surely must restore the primacy of the human bond as artificial intelligence takes on a more prominent role in the diagnosis and practice of medicine. For that to occur, a revamping of how we educate future doctors should start now.

MEDICAL EDUCATION

We select future doctors by their grades in college and the results of the Medical College Admission Test (MCAT). Medical schools began using an entrance exam in the late 1920s when the dropout rate in medical schools in the United States jumped to 50 percent. It was officially named MCAT in 1948, testing science problems and achievement, quantitative ability, and verbal reasoning, with various tweaks over the ensuing decades. A writing sample was part of it for a number of years, but when the most recent iteration was released in 2015 that was abandoned. Now the emphasis is on biological and biochemical systems, biological and psychosocial foundations of behavior, and reasoning skills.

Each year about 20,000 future doctors in the United States are selected (from about 52,000 applicants) on this basis.[58] There is nothing to assess emotional intelligence or the ability to empathize with other human beings. In fact, reliance on metrics of scientific achievement may actually be weeding out the people who are destined to be the most caring individuals, the best communicators, the people who are the most apt to be exemplary healers. By not preparing for current and future technological capabilities, we are setting up the failure of medicine's future to restore humanism.

It makes me think of the recent claim in China that an AI-powered robot—Xiaoyi—passed the national medical licensing examination for the first time. Are we selecting future doctors on a basis that can be simulated or exceeded by an AI bot? I share the view expressed by Joi Ito, who dropped out of college and is now a professor and the head of the MIT Media Lab. Ito said that if there were a system available all of the time that had all the information one needed to memorize to apply for medical school, "maybe there's an argument to be made that you don't have to memorize it." We're certainly headed in that direction. Knowledge, about medicine and individual patients, can and will be outsourced to machine algorithms. What will define and differentiate doctors from their machine apprentices is being human, developing the relationship, witnessing and alleviating suffering. Yes, there will be a need for oversight of the algorithmic output, and that will require science and math reasoning skills. But emotional intelligence needs to take precedence in the selection of future doctors over qualities that are going to be of progressively diminished utility.

Then we get to what happens during medical school. Almost all of the 170 medical and osteopathic schools, with just a couple of exceptions (in an amusing coincidence, the Lerner College of Medicine that I started in Cleveland and the Larner College of Medicine at the University of Vermont), continue to rely on traditional lecturing, rather than moving to innovative active learning that has been proven to improve performance.[59] Most schools do not teach in ways that promote listening and observation skills along with cultivating empathy, despite their proven value in randomized studies.

We also need to rewire the minds of medical students so that they are human-oriented rather than disease-oriented. Hospital and sign-out continuity rounds are all too frequently conducted by "card-flip," whereby one training doctor reviews the disease of the patient, status, and relevant test results without ever going to the patient's bedside. Even the diagnosis of disease is disembodied by our looking at the scan or lab tests instead of laying hands on the person. Such routines are far quicker and easier than getting to know a human being. Rana Awdish, a physician in Detroit, laid this out well with two groups of medical students, one called "pathology" and the other "humanistic." The pathology group gets extraordinary training in recognizing diseases, by recognizing skin lesions, listening for murmurs, or knowing the clotting cascade. The humanistic group gets all that training but also is trained to pursue the context of the human being, letting patients talk and learning about what their lives are like, what is important to them, and what worries them. Given a patient who starts crying, the pathology group can diagnose the disease but can't respond; the humanistic group, wired for emotion even before the tears begin, hears "the tense pitch of vocal cords stretched by false bravery" and comforts. Awdish further writes:

> To disembody doctors and expect them to somehow transcend that handicap and be present in their bodies, empathic and connected is disingenuous. . . . We are not wired to succeed. Medicine cannot heal in a vacuum. It requires connection. . . . We have dedicated our resources to wiring the brains of young doctors to only one way of seeing. They are wired to see disease. They are wired for neglect. But they can be wired for so much more—more depth and beauty and empathy. And everyone, physicians and patients alike, deserves for them to be wired for more.[60]

In *The Lonely Man of Faith*, Rabbi Joseph Soloveitchik interpreted two distinct portrayals of Adam in the early chapters of the book of Genesis. More recently, David Brooks, the *New York Times* columnist, presented an updated description of these two Adams in his book *The Road to Character*. Adam 1 is oriented externally and is ambitious, goal-oriented, and wants to conquer the world. In contrast, Adam 2 is internally anchored, with high moral fiber and a willingness to sacrifice himself to serve others. In many leading medical schools throughout the country, there's an "arms race" for Adam 1s and academic achievement, as Jonathan Stock at Yale University School of Medicine aptly points out.[61] We need to be nurturing the Adam 2s, which is something that is all too often an area of neglect in medical education.

There are many other critical elements that need to be part of the medical school curriculum. Future doctors need a far better understanding of data science, including bioinformatics, biocomputing, probabilistic thinking, and the guts of deep learning neural networks. Much of their efforts in patient care will be supported by algorithms, and they need to understand all the liabilities, to recognize bias, errors, false output, and dissociation from common sense. Likewise, the importance of putting the patient's values and preferences first in any human-machine collaboration cannot be emphasized enough. We can't allow the world of algorithms to propagate medical paternalism, or the suppressive force of doctors to retain control of patient data and medical information, which is long overdue to be terminated (extensively discussed in *The Patient Will See You Now*).[62] Some technology isn't going to be solely AI related, but it still requires rethinking how medicine can be optimally taught—for example, we need to modernize the physical exam if physicians are to routinely incorporate new tools like smartphone ultrasound. Virtual telemedicine will, in many routine circumstances, replace physical visits, and that requires training in "webside manner," which highlights different skills. There's still a face-to-face connection, but, much as skipping a physical exam interferes in the practice of medicine, these physicians will be impaired by an inability to truly connect, lay on hands, and examine the individual, even when better sensors and tools are routinely transferring data remotely. Medical schools are unprepared for these inevitable changes and challenges, unfortunately, because the curriculum is controlled by faculty—the old dogs—who are quick to resist the new machine help that is on the way. The road to deep empathy has to go through revamping medical education. We see the outcry of the new generation, exemplified by Haider Javed Warraich, a Duke Medicine trainee, who wrote, "Young doctors are ready to make health care both more innovative and patient-centric. But are the senior doctors they work with, and the patients they take care of, ready for them?"[63]

DEEP MEDICINE

We're still in the earliest days of AI in medicine. The field is long on computer algorithmic validation and promises but very short on real-world, clinical proof of effectiveness. But with the pace we've seen in just the past few years, with machines outperforming humans on specific, narrow tasks and likely to accelerate and broaden, it is inevitable that narrow AI will take hold. Workflow will improve for most clinicians, be it by faster and more accurate reading of scans and slides, seeing things that humans would miss, or eliminating keyboards so that communication and presence during a clinic visit is restored. At the same time, individuals who so desire will eventually gain the capacity to have their medical data seamlessly aggregated, updated, and processed (along with all the medical literature) to guide them, whether for an optimal diet or their physical or mental health. All of this is surrounded by the caveats that individuals must own and control their medical data, that doctors actively override administrators who desire to sacrifice enhanced human connection in favor of heightened productivity, and that intensive steps to preserve privacy and security of data are taken.

Machine medicine need not be our future. We can choose a technological solution to the profound human disconnection that exists today in healthcare; a more humane medicine, enabled by machine support, can be the path forward. The triad of deep phenotyping— knowing more about the person's layers of medical data than was ever previously attainable or even conceived—deep learning and deep empathy can be a major remedy to the economic crisis in healthcare by promoting bespoke prevention and therapies, superseding many decades of promiscuous and wasteful use of medical resources. But to me, those are the secondary gains of deep medicine. It's our chance, perhaps the ultimate one, to bring back real medicine: Presence. Empathy. Trust. Caring. Being Human.

If you've ever experienced deep pain, you know how lonely and isolating it is, how no one can really know what you are feeling, the anguish, the sense of utter despair. You can be comforted by a loved one, a friend or relative, and that certainly helps. But it's hard to beat the boost from a doctor or clinician you trust and who can bolster your confidence that it will pass, that he or she will be with you no matter what. That you'll be okay. That's the human caring we desperately seek when we're sick. That's what AI can help restore. We may never have another shot like this one. Let's take it.

ACKNOWLEDGMENTS

THIS TURNED OUT to be the most difficult writing project I've ever taken on, for many reasons. Not being a computer scientist, I was fortunate to be able to turn to many people with expertise whom I highly regard for input, including Pedro Domingos, Fei-Fei Li, Gary Marcus, Pearse Keane, Hugh Harvey, Jeremy Howard, Joe Ledsam, and Olaf Ronneberger. Their input was invaluable for providing technical context.

The field of medical AI, while still early, is moving extremely quickly, with something of note coming out every week, and often each day. Reviewing and ingesting all of this material over the past few years represented a formidable challenge, resulting in several hundred references, and I am especially indebted to Michelle Miller at Scripps Research Translational Institute for all her support. My colleagues at Scripps Research, including Steven Steinhubl, Daniel Oran, Emily Spencer, and Giorgio Quer, were very helpful for providing critique.

My editor for all three of the books I've written has been T. J. Kelleher, and I remain appreciative for his insightful input. Similarly,

Katinka Matson, at Brockman, my literary agent for all of my books, has provided stalwart support throughout.

I've been so fortunate and privileged to practice medicine since 1985, after I finished my cardiology training. I have never once lost my love of patient care and appreciate how my patients, in particular, · have been an inspirational force for my efforts to push for a better future of healthcare. With so many of them, I have been privileged to develop treasured relationships that span over thirty years, and I am grateful to all of them for their trust.

I am also indebted to the collaborations that I have formed with industry in different capacities. I have been on the board of directors at Dexcom for several years, an advisor to Illumina, Verily, Walgreens, Blue Cross Blue Shield Association, Quest Diagnostics, and recently Tempus Labs. I do not believe that any of these have influenced my perspective of writing here, but it is important that you know I have potential conflicts of interest. The Scripps Research Translational Institute that I founded in 2006 receives extensive funding from the National Institutes of Health and the Qualcomm Foundation, without which our research would not be possible. I concurrently serve as editor-in-chief for Medscape, a leading medical professional website.

Finally, I want to thank Susan, my wife of forty years, who has supported my hibernating efforts in research and writing, no less for patient care, all through these many decades. She and I are so lucky to have our children, Sarah and Evan, and our grandchildren, Julian and Isabella, just minutes away from our home in La Jolla. They especially make me think a lot about the future, one in which their health will hopefully be far better assured than ours.

NOTES

FOREWORD

1. Broyard, A., *Intoxicated by My Illness.* 2010. New York: Ballantine Books, emphasis mine.
2. Califf, R. M., and R. A. Rosati, "The Doctor and the Computer." *West J Med,* 1981 October. **135**(4): pp. 321–323. https://www.ncbi.nlm.nih.gov/pmc /articles/PMC1273186/.

CHAPTER 1: INTRODUCTION TO DEEP MEDICINE

1. Sisson, P., "Rady Children's Institute Sets Guinness World Record," *San Diego Union Tribune.* 2018.
2. Krizhevsky, A., I. Sutskever, and G. Hinton, "ImageNet Classification with Deep Convolutional Neural Networks," *ACM Digital Library.* 2012: NIPS'12 Proceedings of the 25th International Conference on Neural Information Processing Systems, pp. 1097–1105.
3. Topol, E. J., "Individualized Medicine from Prewomb to Tomb." *Cell,* 2014. **157**(1): pp. 241–253.
4. Schwartz, W. B., "Medicine and the Computer: The Promise and Problems of Change." *N Engl J Med,* 1970. **283**(23): pp. 1257–1264.
5. Peabody, F. W., "The Care of the Patient." *MS/JAMA,* 1927. **88**: pp. 877–882.

CHAPTER 2: SHALLOW MEDICINE

1. Singh, H., A. N. Meyer, and E. J. Thomas, "The Frequency of Diagnostic Errors in Outpatient Care: Estimations from Three Large Observational Studies Involving US Adult Populations." *BMJ Qual Saf,* 2014. **23**(9): pp. 727–731.

2. Cassel, C. K., and J. A. Guest, "Choosing Wisely: Helping Physicians and Patients Make Smart Decisions About Their Care." *JAMA,* 2012. **307**(17): pp. 1801–1802; Mason, D. J., "Choosing Wisely: Changing Clinicians, Patients, or Policies?" *JAMA,* 2015. **313**(7): pp. 657–658; Casarett, D., "The Science of Choosing Wisely—Overcoming the Therapeutic Illusion." *N Engl J Med,* 2016. **374**(13): pp. 1203–1205; "Choosing Wisely: Five Things Physicians and Patients Should Question," *An Initiative of the ABIM Foundation.* American Academy of Allergy & Immunology. 2012.

3. Smith-Bindman, R., "Use of Advanced Imaging Tests and the Not-So-Incidental Harms of Incidental Findings." *JAMA Intern Med,* 2018. **178**(2): pp. 227–228.

4. Casarett, "The Science of Choosing Wisely."

5. Brownlee, S., et al., "Evidence for Overuse of Medical Services Around the World." *Lancet,* 2017. **390**(10090): pp. 156–168; Glasziou, P., et al., "Evidence for Underuse of Effective Medical Services Around the World." *Lancet,* 2017. **390**(10090): pp. 169–177; Saini, V., et al., "Drivers of Poor Medical Care." *Lancet,* 2017. **390**(10090): pp. 178–190; Elshaug, A. G., et al., "Levers for Addressing Medical Underuse and Overuse: Achieving High-Value Health Care." *Lancet,* 2017. **390**(10090): pp. 191–202.

6. Epstein, D., "When Evidence Says No, But Doctors Say Yes," *Atlantic.* February 22, 2017.

7. Bakris, G., and M. Sorrentino, "Redefining Hypertension—Assessing the New Blood-Pressure Guidelines." *N Engl J Med,* 2018. **378**(6): pp. 497–499.

8. Singletary, B., N. Patel, and M. Heslin, "Patient Perceptions About Their Physician in 2 Words: The Good, the Bad, and the Ugly." *JAMA Surg,* 2017. **152**(12): pp. 1169–1170.

9. Brody, B., "Why I Almost Fired My Doctor," *New York Times.* October 12, 2017.

10. Oaklander, M., "Doctors on Life Support," *Time.* 2015.

11. Panagioti, M., et al., "Association Between Physician Burnout and Patient Safety, Professionalism, and Patient Satisfaction: A Systematic Review and Meta-Analysis," *JAMA Intern Med,* 2018.

12. Wang, M. D., R. Khanna, and N. Najafi, "Characterizing the Source of Text in Electronic Health Record Progress Notes." *JAMA Intern Med,* 2017. **177**(8): pp. 1212–1213.

13. Jha, S., "To put this in perspective. Your ATM card works in Outer Mongolia, but your EHR can't be used in a different hospital across the street." Twitter, 2017.

14. Welch, H. G., et al., "Breast-Cancer Tumor Size, Overdiagnosis, and Mammography Screening Effectiveness." *N Engl J Med,* 2016. **375**(15): pp. 1438–1447.

15. "Early Detection of Cancer." Harding Center for Risk Literacy. 2018. https://www.harding-center.mpg.de/en/fact-boxes/early-detection-of-cancer; Pinsky, P. F., P. C. Prorok, and B. S. Kramer, "Prostate Cancer Screening—a Perspective on the Current State of the Evidence." *N Engl J Med,* 2017. **376**(13): pp. 1285–1289; "Prostate-Specific Antigen–Based Screening for Prostate Cancer: A Systematic Evidence Review for the U.S. Preventive Services Task Force," in *Evidence Synthesis Number 154,* 2017.

16. Fraser, M., et al., "Genomic Hallmarks of Localized, Non-Indolent Prostate Cancer." *Nature,* 2017. **541**(7637): pp. 359–364.

17. Pinsky, Prorok, and Kramer, "Prostate Cancer Screening." *N Engl J Med,* 2017.

18. Ahn, H. S., H. J. Kim, and H. G. Welch, "Korea's Thyroid-Cancer 'Epidemic'—Screening and Overdiagnosis." *N Engl J Med,* 2014. **371**(19): pp. 1765–1767.

19. Welch, H. G., "Cancer Screening, Overdiagnosis, and Regulatory Capture." *JAMA Intern Med,* 2017. **177**(7): pp. 915–916.

20. Welch et al., "Breast-Cancer Tumor Size, Overdiagnosis, and Mammography Screening Effectiveness." *N Engl J Med,* 2016. **375**(15), 1438–1447; Welch, "Cancer Screening, Overdiagnosis, and Regulatory Capture."

21. Ghajar, C. M., and M. J. Bissell, "Metastasis: Pathways of Parallel Progression," *Nature.* 2016; Hosseini, H., et al., "Early Dissemination Seeds Metastasis in Breast Cancer," *Nature.* 2016; Townsend, J., "Evolution Research Could Revolutionize Cancer Therapy," *Scientific American.* 2018.

22. Kohane, I. S., Interview with Isaac S. Kohane conducted by Sarah Miller. *Pharmacogenomics,* 2012. **13**(3): pp. 257–260.

23. Welch, "Cancer Screening, Overdiagnosis, and Regulatory Capture."

24. Centers for Medicare and Medicaid Services. August 8, 2018. www.cms.gov/.

25. Silverman, E., "Why Did Prescription Drug Spending Hit $374B in the US Last Year? Read This," *Wall Street Journal.* 2015; Berkrot, B., "U.S. Prescription Drug Spending as High as $610 Billion by 2021: Report," Reuters. 2017.

26. Schork, N. J., "Personalized Medicine: Time for One-Person Trials." *Nature,* 2015. **520**(7549): pp. 609–611.

27. Villarosa, L., "Why America's Black Mothers and Babies Are in a Life-or-Death Crisis," *New York Times.* 2018.

CHAPTER 3: MEDICAL DIAGNOSIS

1. Tversky, A., and D. Kahneman, "Judgment Under Uncertainty: Heuristics and Biases." *Science*, 1974. **185**(4157): pp. 1124–1131.

2. Lewis, M., *The Undoing Project: A Friendship That Changed Our Minds.* 2016. New York: W. W. Norton.

3. Obermeyer, Z., et al., "Early Death After Discharge from Emergency Departments: Analysis of National US Insurance Claims Data." *BMJ*, 2017. **356**: p. j239.

4. Singh, H., A. N. Meyer, and E. J. Thomas, "The Frequency of Diagnostic Errors in Outpatient Care: Estimations from Three Large Observational Studies Involving US Adult Populations." *BMJ Qual Saf*, 2014. **23**(9): pp. 727–731.

5. Brush, J. E., Jr., and J. M. Brophy, "Sharing the Process of Diagnostic Decision Making." *JAMA Intern Med*, 2017. **177**(9): pp. 1245–1246.

6. Tversky and Kahneman, "Judgment Under Uncertainty."

7. Brush and Brophy, "Sharing the Process of Diagnostic Decision Making."

8. "The Internal Medicine Milestone Project," in *The Accreditation Council for Graduate Medical Education and the American Board of Internal Medicine*. 2012.

9. Tetlock, P., *Superforecasting*. 2015. New York: Penguin Random House.

10. Lewis, *The Undoing Project*.

11. Lewis, *The Undoing Project*.

12. Yagoda, B., "The Cognitive Biases Tricking Your Brain," *Atlantic*. 2018.

13. Redelmeier, D. A., and A. Tversky, "Discrepancy Between Medical Decisions for Individual Patients and for Groups." *N Engl J Med*, 1990. **322**(16): pp. 1162–1164.

14. Coussens, S., "Behaving Discretely: Heuristic Thinking in the Emergency Department," *Harvard Scholar*. 2017.

15. Tversky and Kahneman, "Judgment Under Uncertainty."

16. Lewis, *The Undoing Project*.

17. Tversky and Kahneman, "Judgment Under Uncertainty."

18. Topol, E., *The Creative Destruction of Medicine: How the Digital Revolution Will Create Better Health Care*. 2012. New York: Basic Books.

19. Yagoda, "The Cognitive Biases Tricking Your Brain."

20. Yagoda, "The Cognitive Biases Tricking Your Brain."

21. Schiff, G. D., et al., "Diagnostic Error in Medicine: Analysis of 583 Physician-Reported Errors." *Arch Intern Med*, 2009. **169**(20): pp. 1881–1887.

22. Semigran, H. L., et al., "Evaluation of Symptom Checkers for Self Diagnosis and Triage: Audit Study." *BMJ*, 2015. **351**: p. h3480.

23. Van Such, M., et al., "Extent of Diagnostic Agreement Among Medical Referrals." *J Eval Clin Pract,* 2017. **23**(4): pp. 870–874.

24. Muse, E., et al., "From Second to Hundredth Opinion in Medicine: A Global Platform for Physicians." *NPJ Digital Medicine,* in press.

25. Human Diagnosis Project. August 8, 2018. www.humandx.org/.

26. Khazan, O., "Doctors Get Their Own Second Opinions," *Atlantic.* 2017.

27. "Doctor Evidence Brings Valuable Health Data to IBM Watson Ecosystem," IBM Press Release. 2015.

28. Ross, C., and I. Swetlitz, "IBM Pitched Its Watson Supercomputer as a Revolution in Cancer Care: It's Nowhere Close," *Stat News.* 2017.

29. Patel, N. M., et al., "Enhancing Next-Generation Sequencing-Guided Cancer Care Through Cognitive Computing." *Oncologist,* 2018. **23**(2): pp. 179–185.

30. Patel, et al., "Enhancing Next-Generation Sequencing-Guided Cancer Care Through Cognitive Computing."

31. Mukherjee, S., "A.I. Versus M.D.: What Happens When Diagnosis Is Automated?," *New Yorker.* 2017.

32. Ross and Swetlitz, "IBM Pitched Its Watson Supercomputer as a Revolution in Cancer Care."

33. Herper, M., "MD Anderson Benches IBM Watson in Setback for Artificial Intelligence in Medicine," *Forbes.* 2017.

34. Ross and Swetlitz, "IBM Pitched Its Watson Supercomputer as a Revolution in Cancer Care."

35. Muoio, D., "IBM Watson Manager, Academics Describe Challenges, Potential of Health Care AI," *MobiHealthNews.* 2017.

36. Harari, Y. N., *Homo Deus.* 2016. New York: HarperCollins, p. 448.

37. Beam, A. L., and I. S. Kohane, "Translating Artificial Intelligence into Clinical Care." *JAMA,* 2016. **316**(22): pp. 2368–2369.

CHAPTER 4: THE SKINNY ON DEEP LEARNING

1. Dillon, J. J., et al., "Noninvasive Potassium Determination Using a Mathematically Processed ECG: Proof of Concept for a Novel 'Blood-Less,' Blood Test." *J Electrocardiol,* 2015. **48**(1): pp. 12–18.

2. Vic Gundotra, Frank Petterson, and Simon Prakash interview with Eric Topol, *AliveCor.* November 2017.

3. Gundotra, Petterson, and Prakash interview with Topol.

4. Gundotra, Petterson, and Prakash interview with Topol.

5. Comstock, J., "Apple, Stanford Launch Apple Heart Study to Improve Atrial Fibrillation Detection," *MobiHealthNews.* 2017; Loftus, P., and T. Mickle, "Apple Delves Deeper into Health," *Wall Street Journal.* 2017, p. B5.

6. Gonzalez, R., "The New ECG Apple Watch Could Do More Harm Than Good," *Wired*. 2018. https://www.wired.com/story/ecg-apple-watch/; Dormehl, L., "Why We Should Be Wary of Apple Watch 'Ultimate' Health Guardian Claims," Cult of Mac, 2018. https://www.cultofmac.com/577489 /why-we-should-be-wary-of-apple-watch-ultimate-health-guardian-claims/; Victory, J., "What Did Journalists Overlook About the Apple Watch 'Heart Monitor' Feature?" *HealthNewsReview*, 2018. https://www.healthnewsreview .org/2018/09/what-did-journalists-overlook-about-the-apple-watch-heart -monitor-feature/.

7. Goodfellow, I., Y. Bengio, and A. Courville, *Deep Learning*, ed. T. Dietterich. 2016. Cambridge, MA: MIT Press.

8. Domingos, P., *The Master Algorithm*. 2018. New York: Basic Books.

9. Mazzotti, M., "Algorithmic Life," *Los Angeles Review of Books*. 2017.

10. Harari, Y. N., *Homo Deus*. 2016. New York: HarperCollins, p. 348.

11. Harari, *Homo Deus*.

12. Beam, A. L., and I. S. Kohane, "Big Data and Machine Learning in Health Care." *JAMA,* 2018. **319**(13): pp. 1317–1318.

13. Turing, A. M., "On Computable Numbers with an Application to the Entscheidungsproblem." *Proceedings of the London Mathematical Society*, 1936. **42**(1): pp. 230–265. doi: 10.1112/plms/s2-42.1.230.

14. Turing, A. M., "Computing Machinery and Intelligence." *Mind,* 1950. **49**: pp. 433–460. https://www.csee.umbc.edu/courses/471/papers/turing.pdf.

15. Rumelhart, D. E., G. Hinton, and R. J. Williams, "Learning Representations by Back-Propagating Errors." *Nature,* 1986. **323**: pp. 533–536.

16. Parloff, R., "Why Deep Learning Is Suddenly Changing Your Life," in *Fortune*. 2016.

17. Mukherjee, S., "A.I. Versus M.D. What Happens When Diagnosis Is Automated?," *New Yorker*. 2017.

18. Kasparov, G., *Deep Thinking*. vol. 1, 2017. New York: PublicAffairs.

19. Krizhevsky, A., I. Sutskever, and G. Hinton, "ImageNet Classification with Deep Convolutional Neural Networks," *ACM Digital Library*. 2012: NIPS'12 Proceedings of the 25th International Conference on Neural Information Processing Systems, pp. 1097–1105.

20. Esteva, A., et al., "Dermatologist-Level Classification of Skin Cancer with Deep Neural Networks." *Nature,* 2017. **542**(7639): pp. 115–118.

21. Brynjolfsson, E., and T. Mitchell, "What Can Machine Learning Do? Workforce Implications." *Science,* 2017. **358**(6370): pp. 1530–1534.

22. Lin, X., et al., "All-Optical Machine Learning Using Diffractive Deep Neural Networks," *Science*. 2018.

23. LeCun, Y., Y. Bengio, and G. Hinton, "Deep Learning." *Nature*, 2015. **521**(7553): pp. 436–444.

24. Brynjolfsson, E. and T. Mitchell, "What Can Machine Learning Do? Workforce Implications." *Science*, 2017. **358**(6370): pp. 1530–1534.

25. Schaeffer, J., et al., "Checkers Is Solved." *Science,* 2007. **317**(5844): pp. 1518–1522; Sheppard, B., "World-Championship-Caliber Scrabble." *Artificial Intelligence,* 2002. **134**(1–2): pp. 241–275.

26. Mnih, V., et al., "Human-Level Control Through Deep Reinforcement Learning." *Nature,* 2015. **518**.

27. "Why AI Researchers Like Video Games," *Economist.* 2017.

28. Okun, A., and A. Jackson, "Conversations with AlphaGo." *Nature News & Views,* 2017. **550**.

29. Moscovitz, I., "Artificial Intelligence's 'Holy Grail' Victory," *Motley Fool.* 2017.

30. Silver, D., et al., "Mastering the Game of Go with Deep Neural Networks and Tree Search." *Nature,* 2016. **529**(7587): pp. 484–489.

31. Tegmark, M., *Life 3.0: Being Human in the Age of Artificial Intelligence.* 2017. New York: Penguin Random House.

32. Silver, D., et al., "Mastering the Game of Go Without Human Knowledge." *Nature,* 2017. **550**(7676): pp. 354–359.

33. Singh, S., A. Okun, and A. Jackson, "Artificial Intelligence: Learning to Play Go from Scratch." *Nature,* 2017. **550**(7676): pp. 336–337.

34. Silver, D., et al., *Mastering Chess and Shogi by Self-Play with a General Reinforcement Learning Algorithm.* arXiv, 2017.

35. Tegmark, M., "Max Tegmark on Twitter." Twitter, 2017.

36. Bowling, M., et al., "Heads-Up Limit Hold 'Em Poker Is Solved." *Science,* 2015. **347**(6218): pp. 145–149.

37. Moravcik, M., et al., "DeepStack: Expert-Level Artificial Intelligence in Heads-Up No-Limit Poker." *Science,* 2017. **356**(6337): pp. 508–513.

38. Brown, N., and T. Sandholm, "Superhuman AI for Heads-Up No-Limit Poker: Libratus Beats Top Professionals." *Science,* 2017. **359**(6374): pp. 418–424.

39. "Collective Awareness: A Conversation with J. Doyne Farmer," *Edge.* 2018.

40. Markoff, J., "Researchers Announce Advance in Image-Recognition Software," *New York Times.* 2014.

41. Li, F. F., "How We're Teaching Computers to Understand Pictures," *TED.* 2015.

42. Snow, J., "Google's New AI Smile Detector Shows How Embracing Race and Gender Can Reduce Bias," *MIT Technology Review.* 2017.

43. Fowler, G., "Apple Is Sharing Your Face with Apps: That's a New Privacy Worry," *Washington Post.* 2017.

44. Fowler, "Apple Is Sharing Your Face with Apps."

45. Erlich, Y., *Major Flaws in "Identification of Individuals by Trait Prediction Using Whole-Genome Sequencing Data."* arXiv, 2017; Lippert, C., et al., *No Major Flaws in "Identification of Individuals by Trait Prediction Using Whole-Genome Sequencing Data."* arXiv, 2017; Reardon, S., "Geneticists Pan Paper That Claims to Predict a Person's Face from Their DNA," *Nature News & Comment.* 2017.

46. Sheridan, K., "Facial-Recognition Software Finds a New Use: Diagnosing Genetic Disorders," *Stat News.* 2017.

47. Sandoiu, A., "Why Facial Recognition Is the Future of Diagnostics," *Medical News Today.* 2017; Timberg, C., "How Apple Is Bringing Us into the Age of Facial Recognition Whether We're Ready or Not," *Washington Post.* 2017.

48. Hoffman, J., "Reading Pain in a Human Face," *New York Times.* 2014.

49. Nikolov, S., S. Blackwell, R. Mendes, *Deep Learning to Achieve Clinically Applicable Segmentation of Head and Neck Anatomy for Radiotherapy.* arXiv, 2018. https://arxiv.org/abs/1809.04430.

50. Shoham, Y., et al., *Artificial Intelligence Index 2017 Annual Report.* 2017.

51. Upson, S., "The AI Takeover Is Coming: Let's Embrace It," in *Backchannel.* 2016.

52. Lewis-Kraus, G., "The Great A.I. Awakening," *New York Times.* 2016.

53. Knight, W., "An Algorithm Summarizes Lengthy Text Surprisingly Well," *MIT Technology Review.* 2017; Shen, J., et al., *Natural TTS Synthesis by Conditioning WaveNet on Mel Spectrogram Predictions.* arXiv, 2017. **1**.

54. Steinberg, R., "6 Areas Where Artificial Neural Networks Outperform Humans," *Venture Beat.* 2017.

55. Gershgorn, D., "Google's Voice-Generating AI Is Now Indistinguishable from Humans," *Quartz.* 2017.

56. Quain, J. R., "Your Car May Soon Be Able to Read Your Face," *New York Times.* 2017, p. B6.

57. Dixit, V. V., S. Chand, and D. J. Nair, "Autonomous Vehicles: Disengagements, Accidents and Reaction Times." *PLoS One,* 2016. **11**(12): p. e0168054.

58. Halpern, S., "Our Driverless Future," *New York Review of Books.* 2016.

59. Shladover, S., "The Truth About 'Self-Driving' Cars." *Scientific American,* 2016, pp. 53–57.

CHAPTER 5: DEEP LIABILITIES

1. Davis, S. E., T. A. Lasko, G. Chen, E. D. Siew, and M. E. Matheny, "Calibration Drift in Regression and Machine Learning Models for Acute Kidney Injury." *J Am Med Inform Assoc,* 2017. **24**(6): pp. 1052–1061.

2. Chollet, F., *Deep Learning with Python.* 2017. Shelter Island, NY: Manning.

3. Knight, W., "Facebook Heads to Canada for the Next Big AI Breakthrough," *MIT Technology Review.* 2017.

4. Marcus, G., *Deep Learning: A Critical Appraisal.* arXiv, 2018.

5. Hsu, J., "Will the Future of AI Learning Depend More on Nature or Nurture?," in *Spectrum IEEE.* 2017.

6. Rosenfeld, A., R. Zemel, and J. K. Tsotsos, *The Elephant in the Room.* arXiv, 2018. https://arxiv.org/abs/1808.03305.

7. Li, Y., X. Bian, and S. Lyu, *Attacking Object Detectors via Imperceptible Patches on Background.* arXiv, 2018. https://arxiv.org/abs/1809.05966.

8. Somers, J., "Is AI Riding a One-Trick Pony?," *MIT Technology Review.* 2017.

9. Perez, C. E., "Why We Should Be Deeply Suspicious of BackPropagation," *Medium.* 2017.

10. Marcus, *Deep Learning.*

11. Hinton, G., S. Sabour, and N. Frosst, *Matrix Capsules with EM Routing.* 2018. ICLR. Simonite, T., "Google's AI Wizard Unveils a New Twist on Neural Networks," *Wired.* 2017.

12. Silver, D., et al., "Mastering the Game of Go Without Human Knowledge." *Nature,* 2017. **550**(7676): pp. 354–359.

13. Marcus, G., *Gary Marcus Interviews with Eric Topol,* ed. E. Topol. 2017.

14. Collados, J. C., *Is AlphaZero Really a Scientific Breakthrough in AI?* 2017. https://medium.com/@josecamachocollados/is-alphazero-really-a-scientific -breakthrough-in-ai-bf66ae1c84f2.

15. Brouillette, M., "Deep Learning Is a Black Box, but Health Care Won't Mind," *MIT Technology Review.* 2017.

16. Miotto, R., et al., "Deep Patient: An Unsupervised Representation to Predict the Future of Patients from the Electronic Health Records." *Sci Rep,* 2016. **6**: p. 26094.

17. Domingos, P., *Pedro Domingos Interviews with Eric Topol,* ed. E. Topol. 2017.

18. Campolo, A., et al., *AI Now 2017 Report,* ed. S. B. Andrew Selbst. 2017, AI Now Institute.

19. Knight, W., "The Dark Secret at the Heart of AI," *MIT Technology Review.* 2017; Kuang, C., "Can A.I. Be Taught to Explain Itself?" *New York Times.* 2017.

20. Knight, "The Dark Secret at the Heart of AI."

21. Caruana, R., et al., "Intelligible Models for Health Care: Predicting Pneumonia Risk and Hospital 30-Day Readmission," *ACM.* 2015.

22. Kuang, "Can A.I. Be Taught to Explain Itself?"

23. O'Neil, C., *Weapons of Math Destruction: How Big Data Increases Inequality and Threatens Democracy.* 2016. New York: Crown.

24. Zhao, J., et al., *Men Also Like Shopping: Reducing Gender Bias Amplification Using Corpus-Level Constraints.* arXiv, 2017.

25. Simonite, T., "Machines Taught by Photos Learn a Sexist View of Women," *Wired*. 2017.

26. Spice, B., "Questioning the Fairness of Targeting Ads Online," *Carnegie Mellon University News*. 2015.

27. Caliskan, A., J. J. Bryson, and A. Narayanan, "Semantics Derived Automatically from Language Corpora Contain Human-Like Biases." *Science,* 2017. **356**(6334): pp. 183–186.

28. Barr, A., "Google Mistakenly Tags Black People as 'Gorillas,' Showing Limits of Algorithms," *Wall Street Journal*. 2015; Crawford, K., "Artificial Intelligence's White Guy Problem," *New York Times*. 2016.

29. Angwin, J., et al., "Machine Bias," *ProPublica*. 2016.

30. O'Neil, *Weapons of Math Destruction*.

31. Wang, Y., and M. Kosinski, "Deep Neural Networks Are More Accurate Than Humans at Detecting Sexual Orientation from Facial Images." *J Pers Soc Psychol,* 2018. **114**(2): pp. 246–257.

32. Chen, S., "AI Research Is in Desperate Need of an Ethical Watchdog," *Wired*. 2017.

33. Snow, J., "New Research Aims to Solve the Problem of AI Bias in 'Black Box' Algorithms," *MIT Tech Review*. 2017.

34. Snow, "New Research Aims to Solve the Problem of AI Bias in 'Black Box' Algorithms," *MIT Tech Review*. 2017; Tan, S., et al., *Detecting Bias in Black-Box Models Using Transparent Model Distillation*. arXiv, 2017.

35. Crawford, K., "Artificial Intelligence—with Very Real Biases," *Wall Street Journal*. 2017.

36. Vanian, J., "Unmasking A.I.'s Bias Problem," *Fortune*. 2018; Courtland, R., "Bias Detectives: The Researchers Striving to Make Algorithms Fair," *Nature*. 2018.

37. Simonite, T., "Using Artificial Intelligence to Fix Wikipedia's Gender Problem," *Wired*. 2018.

38. Miller, A. P., "Want Less-Biased Decisions? Use Algorithms," *Harvard Business Review*. 2018; Thomas, R., "What HBR Gets Wrong About Algorithms and Bias," *Fast AI*. 2018.

39. Adamson, A. S., and A. Smith, "Machine Learning and Health Care Disparities in Dermatology." *JAMA Dermatol,* 2018.

40. Harari, Y. N., *Homo Deus*. 2016. New York: HarperCollins, p. 348.

41. Lee, K. F., "The Real Threat of Artificial Intelligence," *New York Times*. 2017.

42. Upson, S., "Artificial Intelligence Is Killing the Uncanny Valley and Our Grasp on Reality," *Wired*. 2017.

43. Condliffe, J., "AI Shouldn't Believe Everything It Hears," *MIT Technology Review*. 2017.

44. Cole, S., "AI-Assisted Fake Porn Is Here and We're All Fucked," *Motherboard.* 2017.

45. Suwajanakorn, S., S. M. Seitz, and I. Kemelmacher-Shlizerman, "Synthesizing Obama: Learning Lip Sync from Audio." *ACM Transactions on Graphics,* 2017. **36**(4): pp. 1–13.

46. Knight, W., "Meet the Fake Celebrities Dreamed Up by AI," *MIT Technology Review.* 2017; Karras, T., et al., *Progressive Growing of GANs for Improved Quality, Stability, and Variation.* arXiv, 2017.

47. Erlich, Y., et al., *Re-identification of Genomic Data Using Long Range Familial Searches.* bioRxiv, 2018.

48. Shead, S., "Google DeepMind Has Doubled the Size of Its Healthcare Team," *Business Insider.* 2016; Shead, S., "DeepMind's First Deal with the NHS Has Been Torn Apart in a New Academic Study," *Business Insider.* 2017.

49. Shead, "Google DeepMind Has Doubled the Size of Its Healthcare Team"; Shead, "DeepMind's First Deal with the NHS Has Been Torn Apart in a New Academic Study."

50. Kahn, J., "Alphabet's DeepMind Is Trying to Transform Health Care—but Should an AI Company Have Your Health Records?," *Bloomberg.* 2017.

51. Kahn, J., "Alphabet's DeepMind Is Trying to Transform Health Care."

52. Ibid.

53. Shead, "Google DeepMind Has Doubled the Size of Its Healthcare Team"; Shead, "DeepMind's First Deal with the NHS Has Been Torn Apart in a New Academic Study."

54. Gebru, T., et al., "Using Deep Learning and Google Street View to Estimate the Demographic Makeup of Neighborhoods Across the United States." *Proc Natl Acad Sci USA,* 2017. **114**(50): pp. 13108–13113; Lohr, S., "How Do You Vote? 50 Million Google Images Give a Clue," *New York Times.* 2017.

55. Campolo et al., *AI Now 2017 Report.*

56. Somers, J., "The Coming Software Apocalypse," *Atlantic.* 2017.

57. Papernot, N., and I. Goodfellow, "Privacy and Machine Learning: Two Unexpected Allies?," *cleverhans-blog.* 2018.

58. Etzioni, O., "How to Regulate Artificial Intelligence," *New York Times.* 2017; Simonite, T., "Do We Need a Speedometer for Artificial Intelligence?" *Wired.* 2017.

59. Bonnefon, J. F., A. Shariff, and I. Rahwan, "The Social Dilemma of Autonomous Vehicles." *Science,* 2016. **352**(6293): pp. 1573–1576.

60. Bonnefon, Shariff, and Rahwan, "The Social Dilemma of Autonomous Vehicles."

61. Bonnefon, Shariff, and Rahwan, "The Social Dilemma of Autonomous Vehicles."

62. *Road traffic injuries,* ed. World Health Organization. 2018.

63. Howard, B., "Fatal Arizona Crash: Uber Car Saw Woman, Called It a False Positive," *Extreme Tech.* 2018.

64. *AI for Healthcare: Balancing Efficiency and Ethics,* ed. Infosys. 2017. https://www.infosys.com/smart-automation/Documents/ai-healthcare.pdf.

65. Anthes, E., "The Shape of Work to Come." *Nature,* 2017. **550**(7676): pp. 316–319.

66. Fuhrmans, V., "A Future Without Jobs? Think Again," *Wall Street Journal.* 2017.

67. Kaplan, J., "Don't Fear the Robots," *Wall Street Journal.* 2017.

68. Manyika, J., et al., *Jobs Lost, Jobs Gained: Workforce Transitions in a Time of Automation.* ed. McKinsey Global Institute. 2017. https://www.mckinsey .com/~/media/mckinsey/featured%20insights/future%20of%20organizations /what%20the%20future%20of%20work%20will%20mean%20for%20jobs %20skills%20and%20wages/mgi-jobs-lost-jobs-gained-report-december-6 -2017.ashx.

69. Mason, E. A., "A.I. and Big Data Could Power a New War on Poverty," *New York Times.* 2018.

70. Nedelkoska, L., and G. Quintini, "Automation, Skills Use and Training," in *OECD Social, Employment and Migration Working Papers No. 202.* 2018: OECD, Paris.

71. Gibney, E., "AI Talent Grab Sparks Excitement and Concern." *Nature News & Comment,* 2016. **532**(7600); Metz, C., "N.F.L. Salaries for A.I. Talent," *New York Times.* 2017. pp. B1, B5; Winick, E., "It's Recruiting Season for AI's Top Talent, and Things Are Getting a Little Zany," *MIT Technology Review.* 2017.

72. Etzioni, O., "Workers Displaced by Automation Should Try a New Job: Caregiver," *Wired.* 2017.

73. Pogue, D., "How Well Do Movies Predict Our Tech Future?," *Scientific American.* 2018.

74. Bundy, A., "Smart Machines Are Not a Threat to Humanity." *Communications of the ACM,* 2017. **60**(2): pp. 40–42.

75. Dowd, M., "Elon Musk's Billion-Dollar Crusade to Stop the A.I. Apocalypse," *Vanity Fair.* 2017.

76. *Strategic Plan FY 2014–2018.* HHS Strategic Plan 2017.

77. Dowd, "Elon Musk's Billion-Dollar Crusade to Stop the A.I. Apocalypse"; Russell, S., "Should We Fear Supersmart Robots?," *Scientific American.* 2016, pp. 58–59.

78. Metz, C., "Mark Zuckerberg, Elon Musk and the Feud over Killer Robots," *New York Times.* 2018.

79. Dowd, "Elon Musk's Billion-Dollar Crusade to Stop the A.I. Apocalypse"; Tegmark, M., *Life 3.0: Being Human in the Age of Artificial Intelligence.* 2017. New York: Penguin Random House.

80. Dowd, "Elon Musk's Billion-Dollar Crusade to Stop the A.I. Apocalypse."

81. Dowd, "Elon Musk's Billion-Dollar Crusade to Stop the A.I. Apocalypse."

82. Grace, K., et al., *When Will AI Exceed Human Performance? Evidence from AI Experts.* arXiv, 2017.

83. Khatchadourian, R., "The Doomsday Invention," *New Yorker.* 2015.

84. Tegmark, *Life 3.0.*

CHAPTER 6: DOCTORS AND PATTERNS

1. Jha, S., "Should Radiologists Interact with Patients to Stay Relevant?," *Medscape.* 2017.

2. Wang, X., et al., *ChestX-ray8: Hospital-Scale Chest X-ray Database and Benchmarks on Weakly-Supervised Classification and Localization of Common Thorax Diseases.* arXiv, 2017.

3. Lewis-Kraus, G., "The Great A.I. Awakening," *New York Times.* 2016.

4. Sweeney, E., "Increasingly Powerful AI Systems Are Accompanied by an 'Unanswerable' Question," *FierceHealthcare.* 2017.

5. Rajpurkar, P., et al., *CheXNet: Radiologist-Level Pneumonia Detection on Chest X-Rays with Deep Learning.* arXiv, 2017.

6. Oakden-Rayner, L., "CheXNet: An In-Depth Review," *lukeoakdenrayner.wordpress.com.* 2018.

7. Pachter, L. "When high profile machine learning people oversell their results to the public it leaves everyone else worse off. And how can the public trust scientists if time and time again they are presented with hype instead of science?" Twitter, July 20, 2018. https://twitter.com/lpachter/status/999772391185137664.

8. Jha, S., "Will Computers Replace Radiologists?," *Medscape.* 2016.

9. "Imagine Your World with Watson," *IBM Blog,* 2016.

10. "Mind-Reading Algorithms Reconstruct What You're Seeing Using Brain-Scan Data," *MIT Technology Review.* 2017.

11. Spiegel, A., "Why Even Radiologists Can Miss a Gorilla Hiding in Plain Sight," *Shots—Health News.* 2013.

12. Spiegel, "Why Even Radiologists Can Miss a Gorilla Hiding in Plain Sight."

13. Harvey, H., "Nightmare on ML Street: The Dark Potential of AI in Radiology," *Towards Data Science.* 2017.

14. Yates, E. J., L. C. Yates, and H. Harvey, "Machine Learning 'Red Dot': Open-Source, Cloud, Deep Convolutional Neural Networks in Chest Radiograph Binary Normality Classification." *Clin Radiol*, 2018.

15. Orcutt, M., "Why IBM Just Bought Billions of Medical Images for Watson to Look At," *Technology Review.* 2015.

16. Gillies, R. J., P. E. Kinahan, and H. Hricak, "Radiomics: Images Are More than Pictures, They Are Data." *Radiology,* 2016. **278**(2): pp. 563–577.

17. Akkus, Z., et al., "Predicting Deletion of Chromosomal Arms 1p/19q in Low-Grade Gliomas from MR Images Using Machine Intelligence." *J Digit Imaging,* 2017. **30**(4): pp. 469–476.

18. Ridley, E. L., "Machine Learning Can Help Predict KRAS Mutation Status," *Aunt Minnie.* 2017.

19. Bahl, M., et al., "High-Risk Breast Lesions: A Machine Learning Model to Predict Pathologic Upgrade and Reduce Unnecessary Surgical Excision." *Radiology,* 2018. **286**(3): pp. 810–818.

20. Gale, W., et al., *Detecting Hip Fractures with Radiologist-Level Performance Using Deep Neural Networks.* arXiv, 2017. https://arxiv.org/abs/1711.06504.

21. Sohn, J. H., and T. Vu, "Data-Driven Lung Cancer Risk Stratification of Pulmonary Nodules in Chest CT Using 3D Convolutional Neural Network," in *UCSF Department of Radiology & Biomedical Imaging Symposium.* 2017.

22. Ridley, E. L., "Deep Learning Differentiates Liver Masses on CT," *Aunt Minnie.* 2017.

23. Arbabshirani, M. R., et al., "Advanced Machine Learning in Action: Identification of Intracranial Hemorrhage on Computed Tomography Scans of the Head with Clinical Workflow Integration." *NPJ Digital Medicine,* 2018. **1**(9).

24. Yee, K. M., "AI Algorithm Matches Radiologists in Breast Screening Exams," *Aunt Minnie.* 2017.

25. Ridley, E. L., "Deep Learning Shows Promise for Bone Age Assessment," *Aunt Minnie.* 2017.

26. Nam, J. G., et al., "Development and Validation of a Deep Learning–Based Automated Detection Algorithm for Malignant Pulmonary Nodules on Chest Radiographs." *Radiology,* 2018. https://pubs.rsna.org/doi/10.1148/radiol.2018180237.

27. Bar, A., et al., *Compression Fractures Detection on CT.* arXiv, 2017.

28. Shadmi, R., V. Mazo, and O. Bregman-Amitai, "Fully-Convolutional Deep-Learning Based System for Coronary Calcium Score Prediction from Non-Contrast Chest CT." *IEEE Xplore,* 2018.

29. Idrus, A. A., "Zebra Medical to Offer AI-Based Image Analysis on Google Cloud," *FierceBiotech.* 2017.

30. Siegel, E., "Will Radiologists Be Replaced by Computers? Debunking the Hype of AI," *Carestream.* 2016.

31. Chockley, K., and E. J. Emanuel, "The End of Radiology? Three Threats to the Future Practice of Radiology." *Journal of the American College of Radiology,* 2016. **13**(12): pp. 1415–1420.

32. Ip, G., "How Robots May Make Radiologists' Jobs Easier, Not Redundant," *Wall Street Journal.* 2017.

33. Silverman, L., "Scanning the Future, Radiologists See Their Jobs at Risk," *National Public Radio.* 2017.

34. Grisham, S., "Medscape Physician Compensation Report 2017," *Medscape.* 2017.

35. Bergen, M., "The AI Doctor Orders More Tests," *Bloomberg.* 2017.

36. Bryan, R. N., "Look Ahead—Machine Learning in Radiology," *RSNA News.* 2016.

37. D'Avolio, L., "Thoughts on JAMA's 'Adapting to Artificial Intelligence' by Jha and Topol," LinkedIn. 2017.

38. Recht, M., and R. N. Bryan, "Artificial Intelligence: Threat or Boon to Radiologists?" *J Am Coll Radiol,* 2017. **14**(11): pp. 1476–1480.

39. LeCun, Y., "Disruption in the Workplace: Artificial Intelligence in the 21st Century." YouTube. 2017. https://www.youtube.com/watch?v=OgW4e _ZY26s&t=49s.

40. Silverman, "Scanning the Future."

41. Harvey, H., "Can AI Enable a 10 Minute MRI?," *Towards Data Science.* 2018.

42. Bresnick, J., "Machine Learning 84% Accurate at Flagging Dementia Within 2 Years," *Health IT Analytics.* 2017.

43. Oakden-Rayner, L., et al., "Precision Radiology: Predicting Longevity Using Feature Engineering and Deep Learning Methods in a Radiomics Framework." *Sci Rep,* 2017. **7**(1): p. 1648.

44. Kruskal, J. B., et al., "Big Data and Machine Learning–Strategies for Driving This Bus: A Summary of the 2016 Intersociety Summer Conference." *J Am Coll Radiol,* 2017. **14**(6): pp. 811–817.

45. Levenson, R. M., et al., "Pigeons (*Columba livia*) as Trainable Observers of Pathology and Radiology Breast Cancer Images." *PLoS One,* 2015. **10**(11): p. e0141357.

46. Wang, D., et al., *Deep Learning for Identifying Metastatic Breast Cancer.* arXiv, 2016.

47. Yu, K. H., et al., "Predicting Non–Small Cell Lung Cancer Prognosis by Fully Automated Microscopic Pathology Image Features." *Nat Commun,* 2016. **7**: p. 12474.

48. Hou, L., et al., *Patch-Based Convolutional Neural Network for Whole Slide Tissue Image Classification.* arXiv, 2016.

49. Liu, Y., et al., *Detecting Cancer Metastases on Gigapixel Pathology Images.* arXiv, 2017.

50. Cruz-Roa, A., et al., "Accurate and Reproducible Invasive Breast Cancer Detection in Whole-Slide Images: A Deep Learning Approach for Quantifying Tumor Extent." *Sci Rep,* 2017. **7**: p. 46450.

51. Ehteshami Bejnordi, B., et al., "Diagnostic Assessment of Deep Learning Algorithms for Detection of Lymph Node Metastases in Women with Breast Cancer." *JAMA,* 2017. **318**(22): pp. 2199–2210.

52. Golden, J. A., "Deep Learning Algorithms for Detection of Lymph Node Metastases from Breast Cancer: Helping Artificial Intelligence Be Seen." *JAMA,* 2017. **318**(22): pp. 2184–2186.

53. Yang, S. J., et al., "Assessing Microscope Image Focus Quality with Deep Learning." *BMC Bioinformatics,* 2018. **19**(1): p. 77.

54. Wang et al., *Deep Learning for Identifying Metastatic Breast Cancer.*

55. Wong, D., and S. Yip, "Machine Learning Classifies Cancer." *Nature,* 2018. **555**(7697): pp. 446–447; Capper, D., et al., "DNA Methylation-Based Classification of Central Nervous System Tumours." *Nature,* 2018. **555**(7697): pp. 469–474.

56. Coudray, N., et al., "Classification and Mutation Prediction from Non–Small Cell Lung Cancer Histopathology Images Using Deep Learning." *Nat Med,* 2018.

57. Granter, S. R., A. H. Beck, and D. J. Papke Jr., "AlphaGo, Deep Learning, and the Future of the Human Microscopist." *Arch Pathol Lab Med,* 2017. **141**(5): pp. 619–621.

58. Sharma, G., and A. Carter, "Artificial Intelligence and the Pathologist: Future Frenemies?" *Arch Pathol Lab Med,* 2017. **141**(5): pp. 622–623.

59. Jha, S., and E. J. Topol, "Adapting to Artificial Intelligence: Radiologists and Pathologists as Information Specialists." *JAMA,* 2016. **316**(22): pp. 2353–2354.

60. Patel, N. M., et al., "Enhancing Next-Generation Sequencing-Guided Cancer Care Through Cognitive Computing." *Oncologist,* 2018. **23**(2): pp. 179–185.

61. Wolf, J. A., et al., "Diagnostic Inaccuracy of Smartphone Applications for Melanoma Detection." *JAMA Dermatol,* 2013. **149**(4): pp. 422–426.

62. Resneck, J. S., Jr., et al., "Choice, Transparency, Coordination, and Quality Among Direct-to-Consumer Telemedicine Websites and Apps Treating Skin Disease." *JAMA Dermatol,* 2016. **152**(7): pp. 768–775.

63. Esteva, A., et al., "Dermatologist-Level Classification of Skin Cancer with Deep Neural Networks." *Nature,* 2017. **542**(7639): pp. 115–118.

64. Esteva et al., "Dermatologist-Level Classification of Skin Cancer with Deep Neural Networks."

65. Codella, N., Q. B. Nguyen, and S. Pankanti, *Deep Learning Ensembles for Melanoma Recognition in Dermoscopy Images.* arXiv, 2016.

66. Haenssle, H. A., et al., "Man Against Machine: Diagnostic Performance of a Deep Learning Convolutional Neural Network for Dermoscopic Melanoma Recognition in Comparison to 58 Dermatologists." *Ann Oncol,* 2018.

67. Leachman, S. A., and G. Merlino, "Medicine: The Final Frontier in Cancer Diagnosis." *Nature,* 2017. **542**(7639): pp. 36–38.

68. Esteva et al., "Dermatologist-Level Classification of Skin Cancer with Deep Neural Networks."

69. Zakhem, G. A., C. C. Motosko, and R. S. Ho, "How Should Artificial Intelligence Screen for Skin Cancer and Deliver Diagnostic Predictions to Patients?" *JAMA Dermatol,* 2018.

70. Leswing, K., "Apple CEO Tim Cook Gave a Shout-Out to a $100-per-Year App for Doctors—Here's What It Does," *Business Insider.* 2017.

CHAPTER 7: CLINICIANS WITHOUT PATTERNS

1. Gellert, G., and L. Webster. *The Rise of the Medical Scribe Industry: Implications for Advancement of EHRs,* in *HiMSS 16.* 2016. Las Vegas, NV.

2. Wang, M. D., R. Khanna, and N. Najafi, "Characterizing the Source of Text in Electronic Health Record Progress Notes." *JAMA Intern Med,* 2017. **177**(8): pp. 1212–1213.

3. Bach, B., "Stanford-Google Digital-Scribe Pilot Study to Be Launched," in *Scope.* 2017, Stanford Medicine.

4. Moja, L., et al., "Effectiveness of Computerized Decision Support Systems Linked to Electronic Health Records: A Systematic Review and Meta-Analysis." *Am J Public Health,* 2014. **104**(12): pp. e12–22.

5. Horwitz, R. I., et al., "From Evidence Based Medicine to Medicine Based Evidence." *Am J Med,* 2017. **130**(11): pp. 1246–1250.

6. Lacy, M. E., et al., "Association of Sickle Cell Trait with Hemoglobin A1c in African Americans." *JAMA,* 2017. **317**(5): pp. 507–515.

7. Wong, T. Y., and N. M. Bressler, "Artificial Intelligence with Deep Learning Technology Looks into Diabetic Retinopathy Screening." *JAMA,* 2016. **316**(22): pp. 2366–2367.

8. Wong and Bressler, "Artificial Intelligence with Deep Learning Technology Looks into Diabetic Retinopathy Screening."

9. Gulshan, V., et al., "Development and Validation of a Deep Learning Algorithm for Detection of Diabetic Retinopathy in Retinal Fundus Photographs." *JAMA,* 2016. **316**(22): pp. 2402–2410.

10. Szegedy, C., et al., *Rethinking the Inception Architecture for Computer Vision.* arXiv, 2015.

11. Gulshan et al., "Development and Validation of a Deep Learning Algorithm for Detection of Diabetic Retinopathy in Retinal Fundus Photographs."

12. *IBM Machine Vision Technology Advances Early Detection of Diabetic Eye Disease Using Deep Learning.* 2017.

13. Bleicher, A., "Teenage Whiz Kid Invents an AI System to Diagnose Her Grandfather's Eye Disease." *IEEE Spectrum,* 2017; Lagasse, J., "Teenage Team Develops AI System to Screen for Diabetic Retinopathy," *MobiHealthNews.* 2017.

14. Abramoff, M., et al., "Pivotal Trial of an Autonomous AI-Based Diagnostic System for Detection of Diabetic Retinopathy in Primary Care Offices." *NPJ Digital Medicine,* 2018.

15. Keane, P. and E. Topol, "With an Eye to AI and Autonomous Diagnosis." *NPJ Digital Medicine,* 2018.

16. De Fauw, J., et al., "Clinically Applicable Deep Learning for Diagnosis and Referral in Retinal Disease." *Nature Medicine,* 2018. **24**: pp. 134–1350.

17. Kermany, D. S., et al., "Identifying Medical Diagnoses and Treatable Diseases by Image-Based Deep Learning." *Cell,* 2018. **172**(5): pp. 1122–1131; Rampasek, L., and A. Goldenberg, "Learning from Everyday Images Enables Expert-Like Diagnosis of Retinal Diseases." *Cell,* 2018. **172**(5): pp. 893–895.

18. Poplin, R., et al., "Prediction of Cardiovascular Risk Factors from Retinal Fundus Photographs via Deep Learning." *Nature Biomedical Engineering,* 2018. **2**: pp. 158–164.

19. "The Eye's Structure Holds Information About the Health of the Mind." *Economist.* 2018; Mutlu, U., et al., "Association of Retinal Neurodegeneration on Optical Coherence Tomography with Dementia: A Population-Based Study." *JAMA Neurol,* 2018.

20. Brown, J. M., et al., "Automated Diagnosis of Plus Disease in Retinopathy of Prematurity Using Deep Convolutional Neural Networks." *JAMA Ophthalmol,* 2018. **136**(7): pp. 803–810.

21. Long, E., et al., "An Artificial Intelligence Platform for the Multihospital Collaborative Management of Congenital Cataracts." *Nature Biomedical Engineering,* 2017. **1**: pp. 1–8.

22. Willems, J., et al., "The Diagnostic Performance of Computer Programs for the Interpretation of Electrocardiograms." *NEJM,* 1991. **325**(25): pp. 1767–1773.

23. Heden, B., et al., "Acute Myocardial Infarction Detected in the 12-Lead ECG by Artificial Neural Networks." *Circulation,* 1997. **96**(6): pp. 1798–1802.

24. Heden et al., "Acute Myocardial Infarction Detected in the 12-Lead ECG by Artificial Neural Networks."

25. Strodthoff, N., and C. Strodthoff, *Detecting and Interpreting Myocardial Infarctions Using Fully Convolutional Neural Networks.* arXiv, 2018.

26. Rajpurkar, P., et al., *Cardiologist-Level Arrhythmia Detection with Convolutional Neural Networks.* arXiv, 2017. **1**.

27. Tison, G. H., et al., "Passive Detection of Atrial Fibrillation Using a Commercially Available Smartwatch." *JAMA Cardiol,* 2018. **3**(5): pp. 409–416.

28. Adamson, C., *Ultromics,* ed. E. Topol. 2017.

29. Madani, A., et al., "Fast and Accurate View Classification of Echocardiograms Using Deep Learning." *NPJ Digital Medicine,* 2018. **1**(6).

30. Adamson, *Ultromics.*

31. Le, M., et al., *Computationally Efficient Cardiac Views Projection Using 3D Convolutional Neural Networks.* arXiv, 2017.

32. Weng, S. F., et al., "Can Machine-Learning Improve Cardiovascular Risk Prediction Using Routine Clinical Data?" *PLoS One,* 2017. **12**(4): p. e0174944.

33. Paschalidis, Y., "How Machine Learning Is Helping Us Predict Heart Disease and Diabetes," *Harvard Business Review.* 2017.

34. Manak, M., et al., "Live-Cell Phenotypic-Biomarker Microfluidic Assay for the Risk Stratification of Cancer Patients via Machine Learning." *Nature Biomedical Engineering,* 2018.

35. "Cancer Statistics." National Cancer Institute. July 20, 2018. www .cancer.gov/about-cancer/understanding/statistics.

36. Burns, J., "Artificial Intelligence Is Helping Doctors Find Breast Cancer Risk 30 Times Faster," *Forbes.* 2016.

37. Bahl, M., et al., "High-Risk Breast Lesions: A Machine Learning Model to Predict Pathologic Upgrade and Reduce Unnecessary Surgical Excision." *Radiology,* 2018. **286**(3): pp. 810–818.

38. Lohr, S., "IBM Is Counting on Its Bet on Watson, and Paying Big Money for It," *New York Times.* 2016; Ross, C., "IBM to Congress: Watson Will Transform Health Care, So Keep Your Hands Off Our Supercomputer," *Stat News.* 2017; Mack, H., "IBM Shares Data on How Watson Augments Cancer Treatment Decision-Making," *MobiHealthNews.* 2017; Patel, N. M., et al., "Enhancing Next-Generation Sequencing-Guided Cancer Care Through Cognitive Computing." *Oncologist,* 2018. **23**(2): pp. 179–185; "Watson for Oncology Isn't an AI That Fights Cancer, It's an Unproven Mechanical Turk That Represents the Guesses of a Small Group of Doctors," *Boing Boing.* 2017.

39. Rose, C., "Artificial Intelligence Positioned to Be a Game-Changer," *CBS News.* 2017.

40. Patel et al., "Enhancing Next-Generation Sequencing-Guided Cancer Care Through Cognitive Computing."

41. Patel et al., "Enhancing Next-Generation Sequencing-Guided Cancer Care Through Cognitive Computing."

42. Mack, "IBM Shares Data on How Watson Augments Cancer Treatment Decision-Making."

43. "Watson for Oncology."

44. Ross, C., and I. Swetlitz, "IBM's Watson Supercomputer Recommended 'Unsafe and Incorrect' Cancer Treatments, Internal Documents Show," *Stat News.* 2018; Muller, M., "Playing Doctor: Medical Applications Expose Current Limits of AI," *Spiegel Online.* 2018.

45. McCallister, E., "Computing Care," *Tempus.* 2017.

46. "Tempus Launches New Mobile App to Make Clinical and Genomic Data More Accessible to Physicians at the Point of Care," Associated Press. September 19, 2018. https://www.tempus.com/tempus-launches-new-mobile -app-to-make-clinical-and-genomic-data-more-accessible-to-physicians -at-the-point-of-care/.

47. Versel, N., "Sophia Genetics Looks to Marry Imaging, Genomic Analysis for MDx," *Genome Web.* 2018.

48. Kolata, G., "Colonoscopies Miss Many Cancers, Study Finds," *New York Times.* 2008; Leufkens, A. M., et al., "Factors Influencing the Miss Rate of Polyps in a Back-to-Back Colonoscopy Study." *Endoscopy,* 2012. **44**(5): pp. 470–475.

49. Mori, Y., et al., "Impact of an Automated System for Endocytoscopic Diagnosis of Small Colorectal Lesions: An International Web-Based Study." *Endoscopy,* 2016. **48**(12): pp. 1110–1118; Shin, J. G., et al., "Polyp Missing Rate and Its Associated Risk Factors of Referring Hospitals for Endoscopic Resection of Advanced Colorectal Neoplasia." *Medicine* (Baltimore), 2017. **96**(19): p. e6742.

50. Mori, Y., et al., "Real-Time Use of Artificial Intelligence in Identification of Diminutive Polyps During Colonoscopy." *Annals of Internal Medicine,* 2018. **169**: pp. 357–366.; Holme, O., and L. Aabakken, "Making Colonoscopy Smarter with Standardized Computer-Aided Diagnosis." *Annals of Internal Medicine,* 2018.

51. Mori, Y., et al., "Real-Time Use of Artificial Intelligence in Identification of Diminutive Polyps During Colonoscopy." *Annals of Internal Medicine,* 2018.

52. Aggarwal, A., et al., "Effect of Patient Choice and Hospital Competition on Service Configuration and Technology Adoption Within Cancer Surgery: A National, Population-Based Study." *Lancet Oncol,* 2017. **18**(11): pp. 1445–1453; Abate, C., "Is da Vinci Robotic Surgery a Revolution or a Rip-off?," *Healthline.* 2018.

53. "New Surgical Robots Are About to Enter the Operating Theatre," *Economist.* 2017.

54. Devlin, H., "The Robots Helping NHS Surgeons Perform Better, Faster—and for Longer," *Guardian.* 2018.

55. Taylor, N. P., "After Raising $500M, Fred Moll's Auris Gets FDA Nod for Lung Cancer Robotic Platform," *FierceBiotech.* 2018.

56. Bartolozzi, C., "Neuromorphic Circuits Impart a Sense of Touch." *Science,* 2018. **360**(6392): pp. 966–967.

57. Edwards, T. L., et al., "First-in-Human Study of the Safety and Viability of Intraocular Robotic Surgery." *Nature Biomedical Engineering,* 2018. **2**: pp. 649–656.

58. Huennekens, S., "Surgery 4.0 . . . Digital Surgery 'Democratizing Surgery,'" *Verb Surgical.* 2017.

59. Grace, K., et al., *When Will AI Exceed Human Performance? Evidence from AI Experts.* arXiv, 2017; *The World in 2017, Economist.* 2017.

60. Burton, T., "New Technology Promises to Speed Critical Treatment for Strokes," *Wall Street Journal.* 2018.

61. Titano, J. J., et al., "Automated Deep-Neural-Network Surveillance of Cranial Images for Acute Neurologic Events." *Nat Med,* 2018.

62. Kermany et al., "Identifying Medical Diagnoses and Treatable Diseases by Image-Based Deep Learning."

63. Simon, M., "Tug, the Busy Little Robot Nurse, Will See You Now," *Wired.* 2017.

CHAPTER 8: MENTAL HEALTH

1. "Artificial Intelligence and Psychology: The Computer Will See You Now," *Economist.* 2014.

2. Lucas, G. M., et al., "It's Only a Computer: Virtual Humans Increase Willingness to Disclose." *Computers in Human Behavior,* 2014. **37**: pp. 94–100.

3. Lucas et al., "It's Only a Computer."

4. Lucas et al., "It's Only a Computer."

5. "ELIZA," *Wikipedia.* 2017.

6. Farr, C., "You have an embarrassing medical condition. Would you rather tell and get treatment from: (1) Your doctor; (2) A doctor/nurse; (3) A bot." Twitter, 2017; Knight, W., "Andrew Ng Has a Chatbot That Can Help with Depression," *Technology Review.* 2017.

7. Richardson, J. H., "AI Chatbots Try to Schedule Meetings—Without Enraging Us," *Wired.* 2018.

8. "Podcast: Uncovering the Real Value of AI in Healthcare with Andrew Ng," *Rock Health.* 2017.

9. Insel, T. R., "Digital Phenotyping: Technology for a New Science of Behavior." *JAMA,* 2017. **318**(13): pp. 1215–1216; Or, F., J. Torous, and J. P. Onnela, "High Potential but Limited Evidence: Using Voice Data from Smartphones to Monitor and Diagnose Mood Disorders." *Psychiatr Rehabil J,* 2017. **40**(3): pp. 320–324.

10. Carr, N., "How Smartphones Hijack Our Minds," *Wall Street Journal.* 2017.

11. Nasir, M., et al., "Predicting Couple Therapy Outcomes Based on Speech Acoustic Features." *PLoS One,* 2017. **12**(9): p. e0185123.

12. Bedi, G., et al., "Automated Analysis of Free Speech Predicts Psychosis Onset in High-Risk Youths." *NPJ Schizophr,* 2015. **1**: p. 15030.

13. Frankel, J., "How Artificial Intelligence Could Help Diagnose Mental Disorders," *Atlantic.* 2016.

14. Cao, B., et al., *DeepMood: Modeling Mobile Phone Typing Dynamics for Mood Detection.* arXiv, 2018.

15. Bercovici, J., "Why the Secret to Making Customer Service More Human Isn't Human at All," *Inc. Magazine.* 2017.

16. Stix, C., "3 Ways AI Could Help Our Mental Health," *World Economic Forum.* 2018.

17. Reece, A. G., and C. M. Danforth, "Instagram Photos Reveal Predictive Markers of Depression." *EPJ Data Science,* 2017. **6**.

18. Mitchell, A. J., A. Vaze, and S. Rao, "Clinical Diagnosis of Depression in Primary Care: A Meta-Analysis." *Lancet,* 2009. **374**(9690): pp. 609–619.

19. Landhuis, E., "Brain Imaging Identifies Different Types of Depression," *Scientific American.* 2017.

20. "The Burden of Depression." *Nature,* 2014. **515**(7526): p. 163.

21. Smith, K., "Mental Health: A World of Depression." *Nature,* 2014. **515**(7526): p. 181.

22. McConnon, A., "AI-Powered Systems Target Mental Health," *Wall Street Journal.* 2018.

23. Winick, E., "With Brain-Scanning Hats, China Signals It Has No Interest in Workers' Privacy," *MIT Technology Review.* 2018.

24. Schnyer, D. M., et al., "Evaluating the Diagnostic Utility of Applying a Machine Learning Algorithm to Diffusion Tensor MRI Measures in Individuals with Major Depressive Disorder." *Psychiatry Res,* 2017. **264**: pp. 1–9.

25. Schnyer et al., "Evaluating the Diagnostic Utility." Drysdale, A. T., et al., "Resting-State Connectivity Biomarkers Define Neurophysiological Subtypes of Depression." *Nat Med,* 2017. **23**(1): pp. 28–38.

26. Schnyer et al., "Evaluating the Diagnostic Utility."

27. Comstock, J., "Sonde Health Will Use MIT Voice Analysis Tech to Detect Mental Health Conditions," *MobiHealthNews.* 2016.

28. Vergyri, D., et al., "Speech-Based Assessment of PTSD in a Military Population Using Diverse Feature Classes." *Proc. Interspeech,* 2015: pp. 3729–3733.

29. Scherer, S., et al., "Self-Reported Symptoms of Depression and PTSD Are Associated with Reduced Vowel Space in Screening Interviews." *IEEE Transactions on Affective Computing,* 2015. **7**(1): pp. 59–73.

30. Or, Torous, and Onnela, "High Potential but Limited Evidence."

31. Chekroud, A. M., et al., "Cross-Trial Prediction of Treatment Outcome in Depression: A Machine Learning Approach." *Lancet Psychiatry,* 2016. **3**(3): pp. 243–250.

32. Hutson, M., "Machine-Learning Algorithms Can Predict Suicide Risk More Readily Than Clinicians, Study Finds," *Newsweek.* 2017.

33. "Suicide Statistics," American Foundation for Suicide Prevention, July 19, 2018. https://afsp.org/about-suicide/suicide-statistics/.

34. Denworth, L., "Could a Machine Identify Suicidal Thoughts?," *Scientific American.* 2017.

35. Franklin, J. C., et al., "Risk Factors for Suicidal Thoughts and Behaviors: A Meta-Analysis of 50 Years of Research." *Psychol Bull,* 2017. **143**(2): pp. 187–232; McConnon, A., "AI Helps Identify Those at Risk for Suicide," *Wall Street Journal.* 2018. p. R7.

36. Franklin et al., "Risk Factors for Suicidal Thoughts and Behaviors."

37. Walsh, C. G., et al., "Predicting Risk of Suicide Attempts over Time Through Machine Learning." *Clinical Psychological Science,* 2017. **5**(3): pp. 457–469.

38. Hutson, "Machine-Learning Algorithms Can Predict Suicide Risk." Walsh et al., "Predicting Risk of Suicide Attempts."

39. Hutson, "Machine-Learning Algorithms Can Predict Suicide Risk."

40. Hutson, "Machine-Learning Algorithms Can Predict Suicide Risk"; Horwitz, B., "Identifying Suicidal Young Adults." *Nature Human Behavior,* 2017. **1**: pp. 860–861.

41. Cheng, Q., et al., "Assessing Suicide Risk and Emotional Distress in Chinese Social Media: A Text Mining and Machine Learning Study." *J Med Internet Res,* 2017. **19**(7): p. e243.

42. McConnon, "AI Helps Identify Those at Risk for Suicide."

43. "Crisis Trends," July 19, 2018. https://crisistrends.org/#visualizations.

44. Resnick, B., "How Data Scientists Are Using AI for Suicide Prevention," *Vox.* 2018.

45. Anthes, E., "Depression: A Change of Mind." *Nature,* 2014. **515**(7526): pp. 185–187.

46. Firth, J., et al., "The Efficacy of Smartphone-Based Mental Health Interventions for Depressive Symptoms: A Meta-Analysis of Randomized Controlled Trials." *World Psychiatry,* 2017. **16**(3): pp. 287–298.

47. Aggarwal, J., and W. Smriti Joshi, "The Future of Artificial Intelligence in Mental Health," DQINDIA online. 2017.

48. Fitzpatrick, K. K., A. Darcy, and M. Vierhile, "Delivering Cognitive Behavior Therapy to Young Adults with Symptoms of Depression and Anxiety Using a Fully Automated Conversational Agent (Woebot): A Randomized Controlled Trial." *JMIR Ment Health,* 2017. **4**(2): p. e19.

49. Knight, "Andrew Ng Has a Chatbot That Can Help with Depression."

50. Lien, T., "Depressed but Can't See a Therapist? This Chatbot Could Help," *Los Angeles Times.* 2017.

51. Lien, "Depressed but Can't See a Therapist?"

52. Ben-Zeev, D., and D. C. Atkins, "Bringing Digital Mental Health to Where It Is Needed Most." *Nature Human Behavior,* 2017. **1**: pp. 849–851;

Barrett, P. M., et al., "Digitising the Mind." *Lancet,* 2017. **389**(10082): p. 1877.

53. Nutt, A. E., "'The Woebot Will See You Now'—the Rise of Chatbot Therapy," *Washington Post.* 2017.

54. Smith, "Mental Health."

55. Romeo, N., "The Chatbot Will See You Now," *New Yorker.* 2016.

56. Fitzpatrick, Darcy, and Vierhile, "Delivering Cognitive Behavior Therapy."

57. Pugh, A., "Automated Health Care Offers Freedom from Shame, but Is It What Patients Need?" *New Yorker.* 2018.

58. Harari, Y. N., *Homo Deus.* 2016. New York: HarperCollins, p. 448.

59. Budner, P., J. Eirich, and P. A. Gloor, *"Making You Happy Makes Me Happy": Measuring Individual Mood with Smartwatches.* arXiv, 2017. arXiv:1711.06134 [cs.HC]. "How a Smart Watch Can Predict Your Happiness Levels," *MIT Technology Review.* 2017.

60. Clark, A. E., et al., "The Key Determinants of Happiness and Misery," *World Happiness Report.* 2017; "Daily Chart: A New Study Tries to Unpick What Makes People Happy and Sad," *Economist.* 2017.

61. Hwang, J. J., et al., *Learning Beyond Human Expertise with Generative Models for Dental Restorations.* arXiv, 2018.

62. Peters, A., "Having a Heart Attack? This AI Helps Emergency Dispatchers Find Out," *Fast Company.* 2018.

CHAPTER 9: AI AND HEALTH SYSTEMS

1. Avati, A., et al., *Improving Palliative Care with Deep Learning.* arXiv, 2017; Mukherjee, S., "This Cat Sensed Death: What If Computers Could, Too?," *New York Times.* 2018; Bergen, M., "Google Is Training Machines to Predict When a Patient Will Die," *Bloomberg.* 2018.

2. Avati et al., *Improving Palliative Care with Deep Learning;* Snow, J., "A New Algorithm Identifies Candidates for Palliative Care by Predicting When Patients Will Die," *MIT Technology Review.* 2017; White, N., et al., "A Systematic Review of Predictions of Survival in Palliative Care: How Accurate Are Clinicians and Who Are the Experts?" *PLoS One,* 2016. **11**(8): p. e0161407.

3. Bennington-Castro, J., "A New Algorithm Could Ease Critically Ill Patients' Final Days," *NBC News.* 2018.

4. White, N., et al., "How Accurate Is the 'Surprise Question' at Identifying Patients at the End of Life? A Systematic Review and Meta-Analysis." *BMC Med,* 2017. **15**(1): p. 139.

5. Avati et al., *Improving Palliative Care with Deep Learning;* Mukherjee, "This Cat Sensed Death."

6. Zaidi, D., "AI Is Transforming Medical Diagnosis, Prosthetics, and Vision Aids," *Venture Beat.* 2017.

7. Rajkomar, A., et al., "Scalable and Accurate Deep Learning with Electronic Health Records." *NPJ Digital Medicine,* 2018.

8. Meyer, A., et al., "Real-Time Prediction of Death, Renal Failure and Postoperative Bleeding in Post-Cardiothoracic Critical Care Using Deep Learning on Routinely Collected Clinical Data." *Lancet,* in press.

9. Mullin, E., "DeepMind's New Project Aims to Prevent Hospital Deaths," *MIT Technology Review.* 2018.

10. Yoon, J., et al., "Personalized Survival Predictions via Trees of Predictors: An Application to Cardiac Transplantation." *PLoS One,* 2018. **13**(3): p. e0194985.

11. Son, J. H., et al., "Deep Phenotyping on Electronic Health Records Facilitates Genetic Diagnosis by Clinical Exomes." *Am J Hum Genet,* 2018. **103**(1): pp. 58–73.

12. Mukherjee, "This Cat Sensed Death."

13. O'Neil, C., "Big Data Is Coming to Take Your Health Insurance," *Bloomberg.* 2017; Gillin, P., "How Machine Learning Will Spark a Revolution in Insurance," *Silicon Angle.* 2017; Lecher, C., "What Happens When an Algorithm Cuts Your Health Care," *Verge.* 2018.

14. Ross, C., "The Data Are In, but Debate Rages: Are Hospital Readmission Penalties a Good Idea?," *Stat News.* 2017.

15. Shameer, K., et al., "Predictive Modeling of Hospital Readmission Rates Using Electronic Medical Record-Wide Machine Learning: A Case-Study Using Mount Sinai Heart Failure Cohort." *Pac Symp Biocomput,* 2017. **22**: pp. 276–287.

16. Nguyen, P., et al., "Deepr: A Convolutional Net for Medical Records." *IEEE J Biomed Health Inform,* 2017. **21**(1): pp. 22–30.

17. Choi, E., et al., "Doctor AI: Predicting Clinical Events via Recurrent Neural Networks." *JMLR Workshop Conf Proc,* 2016. **56**: pp. 301–318.

18. Yang, Z., et al., "Clinical Assistant Diagnosis for Electronic Medical Record Based on Convolutional Neural Network." *Sci Rep,* 2018. **8**(1): p. 6329.

19. Razavian, N., J. Marcus, and D. Sontag, "Multi-Task Prediction of Disease Onsets from Longitudinal Lab Tests." *PMLR,* 2016. **56**: pp. 73–100.

20. Avati et al., *Improving Palliative Care with Deep Learning;* Rajkomar et al., "Scalable and Accurate Deep Learning with Electronic Health Records"; Shameer et al., "Predictive Modeling of Hospital Readmission Rates"; Yang, Z., et al., "Clinical Assistant Diagnosis for Electronic Medical Record Based on Convolutional Neural Network." *Sci Rep,* 2018. **8**(1): p. 6329; Razavian, Marcus, and Sontag, "Multi-task Prediction of Disease Onsets"; Oh, J., et al., "A Generalizable, Data-Driven Approach to Predict Daily Risk of Clostridium Difficile Infection at Two Large Academic Health Centers." *Infect Control Hosp Epidemiol,* 2018. **39**(4): pp. 425–433; Miotto, R., et al., "Deep Patient:

An Unsupervised Representation to Predict the Future of Patients from the Electronic Health Records." *Sci Rep,* 2016. **6**: p. 26094; Mathotaarachchi, S., et al., "Identifying Incipient Dementia Individuals Using Machine Learning and Amyloid Imaging." *Neurobiol Aging,* 2017. **59**: pp. 80–90; Elfiky, A., et al., "Development and Application of a Machine Learning Approach to Assess Short-Term Mortality Risk Among Patients with Cancer Starting Chemotherapy." *JAMA Network Open,* 2018; Horng, S., et al., "Creating an Automated Trigger for Sepsis Clinical Decision Support at Emergency Department Triage Using Machine Learning." *PLoS One,* 2017. **12**(4): p. e0174708; Walsh, C. G., et al., "Predicting Risk of Suicide Attempts over Time Through Machine Learning." *Clinical Psychological Science,* 2017. **5**(3): pp. 457–469; Wong, A., et al., "Development and Validation of an Electronic Health Record–Based Machine Learning Model to Estimate Delirium Risk in Newly Hospitalized Patients Without Known Cognitive Impairment." *JAMA Network Open,* 2018; Henry, K. E., et al., "A Targeted Real-Time Early Warning Score (TREWScore) for Septic Shock." *Sci Transl Med,* 2015. **7**(299): p. 299ra122; Culliton, P., et al., *Predicting Severe Sepsis Using Text from the Electronic Health Record.* arXiv, 2017; Cleret de Langavant, L., E. Bayen, and K. Yaffe, "Unsupervised Machine Learning to Identify High Likelihood of Dementia in Population-Based Surveys: Development and Validation Study." *J Med Internet Res,* 2018. **20**(7): p. e10493.

21. *Current Employment Statistics Highlights,* ed. N. E. Branch. 2018, US Bureau of Labor Statistics.

22. Terhune, C., "Our Costly Addiction to Health Care Jobs," *New York Times.* 2017.

23. Terhune, "Our Costly Addiction to Health Care Jobs."

24. Lee, K. F., "Tech Companies Should Stop Pretending AI Won't Destroy Jobs," *MIT Technology Review.* 2018.

25. Tseng, P., et al., "Administrative Costs Associated with Physician Billing and Insurance-Related Activities at an Academic Health Care System." *JAMA,* 2018. **319**(7): pp. 691–697.

26. Frakt, A., "The Astonishingly High Administrative Costs of U.S. Health Care," *New York Times.* 2018.

27. InoviaGroup, *Artificial Intelligence Virtual Assist (AIVA).* August 9, 2018. http://inoviagroup.se/artificial-intelligence-virtual-assist-aiva/.

28. Muoio, D., "Qventus Receives $30M Investment to Bring AI to Hospital Workflows," *MobiHealthNews.* 2018.

29. Zweig, M., D. Tran, and B. Evans, "Demystifying AI and Machine Learning in Healthcare," *Rock Health.* 2018; Ockerman, E., "AI Hospital Software Knows Who's Going to Fall," *Bloomberg Businessweek.* 2018.

30. Siwicki, B., "Radiology Practices Using AI and NLP to Boost MIPS Payments," *Healthcare IT News.* 2018.

31. Sohn, E., et al., "Four Lessons in the Adoption of Machine Learning in Health Care," *Health Affairs.* 2017.

32. Zhu, B., et al., "Image Reconstruction by Domain-Transform Manifold Learning." *Nature,* 2018. **555**(7697): pp. 487–492; Harvey, H., "Can AI Enable a 10 Minute MRI?," *Towards Data Science.* 2018. Ridley, E. L., "Artificial Intelligence Guides Lower PET Tracer Dose." *Aunt Minnie,* 2018.

33. Nikolov, S., S. Blackwell, R. Mendes, *Deep Learning to Achieve Clinically Applicable Segmentation of Head and Neck Anatomy for Radiotherapy.* arXiv, 2018. https://arxiv.org/abs/1809.04430.

34. Henry, K. E., "A Targeted Real-Time Early Warning Score (TREW-Score) for Septic Shock"; Liu, V. X., and A. J. Walkey, "Machine Learning and Sepsis: On the Road to Revolution." *Crit Care Med,* 2017. **45**(11): pp. 1946–1947; Horng et al., "Creating an Automated Trigger for Sepsis Clinical Decision Support"; Chan, R., "A.I. Can Predict Whether You Have Sepsis Before Doctors Even Know It," *Inverse.* 2017; Nemati, S., et al., "An Interpretable Machine Learning Model for Accurate Prediction of Sepsis in the ICU." *Crit Care Med,* 2017.

35. McQuaid, J., "To Fight Fatal Infections, Hospitals May Turn to Algorithms," *Scientific American.* 2018.

36. Oh et al., "A Generalizable, Data-Driven Approach to Predict Daily Risk of Clostridium Difficile."

37. Haque, A., et al., *Towards Vision-Based Smart Hospitals: A System for Tracking and Monitoring Hand Hygiene Compliance.* arXiv, 2017. Yeung, S., et al., "Bedside Computer Vision—Moving Artificial Intelligence from Driver Assistance to Patient Safety." *N Engl J Med,* 2018. **378**(14): pp. 1271–1273.

38. Prasad, N., L. F. Cheng, C. Chivers, M. Draugelis, and B. E. Engelhardt, *A Reinforcement Learning Approach to Weaning of Mechanical Ventilation in Intensive Care Units.* arXiv, 2017. https://arxiv.org/abs/1704.06300.

39. Suresh, H., et al., *Clinical Intervention Prediction and Understanding with Deep Neural Networks.* arXiv, 2017.

40. Gordon, R., "Using Machine Learning to Improve Patient Care," *MIT News.* 2017.

41. Maier-Hein, L., et al., "Surgical Data Science for Next-Generation Interventions." *Nature Biomedical Engineering,* 2017. **1**: pp. 691–696.

42. "Artificial Intelligence Will Improve Medical Treatments," *Economist.* 2018.

43. Burton, T., "New Stroke Technology to Identify Worst Cases Gets FDA Approval," *Wall Street Journal.* 2018.

44. Auerbach, D. I., D. O. Staiger, and P. I. Buerhaus, "Growing Ranks of Advanced Practice Clinicians—Implications for the Physician Workforce." *N Engl J Med,* 2018. **378**(25): pp. 2358–2360.

45. Libberton, B., "Career Advice and an Inside Perspective on Being a Researcher," *Karolinska Institute Career Blog.* 2017.

46. Hu, J., "A Hospital Without Patients," *Politico.* 2017.

47. Zhu et al., "Image Reconstruction by Domain-Transform Manifold Learning."

48. Kwolek, B., and M. Kepski, "Human Fall Detection on Embedded Platform Using Depth Maps and Wireless Accelerometer." *Comput Methods Programs Biomed,* 2014. **117**(3): pp. 489–501; Billis, A. S., et al., "A Decision-Support Framework for Promoting Independent Living and Ageing Well." *IEEE J Biomed Health Inform,* 2015. **19**(1): pp. 199–209; Press, G., "A New AI-Driven Companion for Older Adults, Improving Their Quality of Life," *Forbes.* 2017.

49. Kodjak, A., and S. Davis, "Trump Administration Move Imperils Pre-Existing Condition Protections," *NPR.* 2018.

50. Madison, K., "The Risks of Using Workplace Wellness Programs to Foster a Culture of Health" in *Health Affairs,* 2016. **35**(11): pp. 2068–2074.

51. Taddeo, M., and L. Floridi, "Regulate Artificial Intelligence to Avert Cyber Arms Race." *Nature,* 2018. **556**(7701): pp. 296–298.

52. Onstad, K., "The AI Superstars at Google, Facebook, Apple—They All Studied Under This Guy: Mr. Robot," *Toronto Life.* 2018.

53. Deshpande, P., "AI Could Help Solve the World's Health Care Problems at Scale," *Venture Beat.* 2017.

54. "China May Match or Beat America in AI," *Economist.* 2017; Bremmer, I., "China Embraces AI: A Close Look and A Long View," *Sinovation Ventures,* ed. E. Group. 2017; Zhang, S., "China's Artificial-Intelligence Boom," *Atlantic.* 2017; Lin, L., "Facial Recognition Wears a Smile," *Wall Street Journal.* 2017; "Who Is Winning the AI Race?," *MIT Technology Review.* 2017.

55. Wee, S. L., "China's Tech Titans, Making Gains in A.I., Improve Health Care," *New York Times.* 2018. p. B7.

56. Wee, S. L., "China's Tech Titans."

57. Metz, C., "As China Marches Forward on A.I., the White House Is Silent," *New York Times.* 2018.

58. Larson, C., "China's AI Imperative." *Science,* 2018. **359**(6376): pp. 628–630.

59. Huang, E., "A Chinese Hospital Is Betting Big on Artificial Intelligence to Treat Patients," *Quartz.* 2018.

60. Galeon, D., "For the First Time, a Robot Passed a Medical Licensing Exam," *Futurism.* 2017; Si, M., and C. Yu, "Chinese Robot Becomes World's First Machine to Pass Medical Exam," *China Daily.* 2017.

61. Sun, Y., "AI Could Alleviate China's Doctor Shortage," *MIT Technology Review.* 2018.

62. Knight, W., "Meet the Chinese Finance Giant That's Secretly an AI Company," *MIT Technology Review.* 2017.

63. Millward, J. A., "What It's Like to Live in a Surveillance State," *New York Times*. 2018.

64. Villani, C., *For a Meaningful Artificial Intelligence*. ed. AI for Humanity. 2018.

65. Thompson, N., "Emmanuel Macron Q&A: France's President Discusses Artificial Intelligence Strategy," *Wired*. 2018.

66. Perkins, A., "May to Pledge Millions to AI Research Assisting Early Cancer Diagnosis," *Guardian*. 2018.

67. *The Topol Review*. 2018, NHS Health Education England. www.hee .nhs.uk/our-work/topol-review.

CHAPTER 10: DEEP DISCOVERY

1. Camacho, D. M., et al., "Next-Generation Machine Learning for Biological Networks." *Cell*, 2018. **173**(7): pp. 1581–1592.

2. Appenzeller, T., "The Scientists' Apprentice." *Science Magazine*, 2017. **357**(6346): pp. 16–17.

3. Zhou, J., and O. G. Troyanskaya, "Predicting Effects of Noncoding Variants with Deep Learning–Based Sequence Model." *Nat Methods*, 2015. **12**(10): pp. 931–934; Pennisi, E., "AI in Action: Combing the Genome for the Roots of Autism." *Science*, 2017. **357**(6346): p. 25.

4. Krishnan, A., et al., "Genome-Wide Prediction and Functional Characterization of the Genetic Basis of Autism Spectrum Disorder." *Nat Neurosci*, 2016. **19**(11): pp. 1454–1462.

5. Molteni, M., "Google Is Giving Away AI That Can Build Your Genome Sequence," *Wired*. 2017; Carroll, A. and N. Thangaraj, "Evaluating DeepVariant: A New Deep Learning Variant Caller from the Google Brain Team," *DNA Nexus*. 2017; Poplin, R., et al., *Creating a Universal SNP and Small Indel Variant Caller with Deep Neural Networks*. bioRxiv, 2016; DePristo, M., and R. Poplin, "DeepVariant: Highly Accurate Genomes with Deep Neural Networks," *Google Research Blog*. 2017.

6. Zhou, J., et al., "Deep Learning Sequence–Based Ab Initio Prediction of Variant Effects on Expression and Disease Risk." *Nat Genet*, 2018. **50**(8): pp. 1171–1179.

7. Sundaram, L., et al., "Predicting the Clinical Impact of Human Mutation with Deep Neural Networks." *Nat Genet*, 2018. **50**(8): pp. 1161–1170.

8. Camacho et al., "Next-Generation Machine Learning for Biological Networks"; Ching, T., et al., *Opportunities and Obstacles for Deep Learning in Biology and Medicine*. bioRxiv, 2017; AlQuraishi, M., *End-to-End Differentiable Learning of Protein Structure*. bioRxiv, 2018; Zitnik, M., et al., *Machine Learning for Integrating Data in Biology and Medicine: Principles, Practice, and Opportunities*. arXiv, 2018.

9. Riesselman, A., J. Ingraham, and D. Marks, "Deep Generative Models of Genetic Variation Capture the Effects of Mutations." *Nature Methods,* 2018; Poplin, R., et al., "A Universal SNP and Small-Indel Variant Caller Using Deep Neural Networks." *Nat Biotechnol,* 2018.

10. Miotto, R., et al., "Deep Learning for Healthcare: Review, Opportunities and Challenges." *Brief Bioinform,* 2017. https://www.ncbi.nlm.nih.gov/pubmed/28481991.

11. Angermueller, C., et al., "DeepCpG: Accurate Prediction of Single-Cell DNA Methylation States Using Deep Learning." *Genome Biol,* 2017. **18**(1): p. 67.

12. Miotto et al., "Deep Learning for Healthcare."

13. Lin, C., et al., "Using Neural Networks for Reducing the Dimensions of Single-Cell RNA-Seq Data." *Nucleic Acids Res,* 2017. **45**(17): p. e156.

14. van Dijk, D., et al., "Recovering Gene Interactions from Single-Cell Data Using Data Diffusion." *Cell,* 2018. **174**(3): pp. 716–729 e27.

15. LeFebvre, R., "Microsoft AI Is Being Used to Improve CRISPR Accuracy," *Engadget.* 2018; Listgarten, J., et al., "Prediction of Off-Target Activities for the End-to-End Design of CRISPR Guide RNAs." *Nature Biomedical Engineering,* 2018. **2**: pp. 38–47.

16. Buggenthin, F., et al., "Prospective Identification of Hematopoietic Lineage Choice by Deep Learning." *Nat Methods,* 2017. **14**(4): pp. 403–406; Webb, S., "Deep Learning for Biology." *Nature,* 2018. **554**(7693): pp. 555–557.

17. Ma, J., et al., "Using Deep Learning to Model the Hierarchical Structure and Function of a Cell." *Nat Methods,* 2018. **15**(4): pp. 290–298.

18. Wrzeszczynski, K. O., et al., "Comparing Sequencing Assays and Human-Machine Analyses in Actionable Genomics for Glioblastoma." *Neurol Genet,* 2017. **3**(4): pp. e164.

19. Wong, D., and S. Yip, "Machine Learning Classifies Cancer." *Nature,* 2018. **555**(7697): pp. 446–447; Capper, D., et al., "DNA Methylation–Based Classification of Central Nervous System Tumours." *Nature,* 2018. **555**(7697): pp. 469–474.

20. Caravagna, G., Y. Giarratano, D. Ramazzotti, I. Tomlinson, et al., "Detecting Repeated Cancer Evolution from Multi-Region Tumor Sequencing Data." *Nature Methods,* 2018. **15**: pp. 707–714.

21. Sheldrick, G., "Robot War on Cancer: Scientists Develop Breakthrough AI Tech to Predict How Tumours Grow." Express.co.uk. 2018.

22. Wood, D.E., et al., "A Machine Learning Approach for Somatic Mutation Discovery." *Sci Transl Med,* 2018. **10**(457).

23. Behravan, H., et al., "Machine Learning Identifies Interacting Genetic Variants Contributing to Breast Cancer Risk: A Case Study in Finnish Cases and Controls." *Sci Rep,* 2018. **8**(1): p. 13149.

24. Lobo, D., M. Lobikin, and M. Levin, "Discovering Novel Phenotypes with Automatically Inferred Dynamic Models: A Partial Melanocyte Conversion in Xenopus." *Sci Rep,* 2017. **7**: p. 41339.

25. Nelson, B., "Artificial Intelligence Could Drastically Reduce the Time It Takes to Develop New Life-Saving Drugs," *NBC News MACH.* 2018.

26. Zainzinger, V., "New Digital Chemical Screening Tool Could Help Eliminate Animal Testing," *Science Magazine.* 2018.

27. Mullard, A., "The Drug-Maker's Guide to the Galaxy." *Nature,* 2017. **549**(7673): pp. 445–447.

28. Mullard, "The Drug-Maker's Guide to the Galaxy."

29. Service, R. F., "AI in Action: Neural Networks Learn the Art of Chemical Synthesis." *Science,* 2017. **357**(6346): p. 27.

30. Bilsland, E., et al., "Plasmodium Dihydrofolate Reductase Is a Second Enzyme Target for the Antimalarial Action of Triclosan." *Sci Rep,* 2018. **8**(1): p. 1038.

31. Ahneman, D. T., et al., "Predicting Reaction Performance in C-N Cross-Coupling Using Machine Learning." *Science,* 2018. **360**(6385): pp. 186–190.

32. Dilawar, A., "The Artificial Miracle," *PressReader.* 2017.

33. Segler, M. H. S., M. Preuss, and M. P. Waller, "Planning Chemical Syntheses with Deep Neural Networks and Symbolic AI." *Nature,* 2018. **555**(7698): pp. 604–610.

34. Else, H., "Need to Make a Molecule? Ask This AI for Instructions." *Nature,* 2018.

35. Granda, J. M., et al., "Controlling an Organic Synthesis Robot with Machine Learning to Search for New Reactivity." *Nature,* 2018. **559**(7714): pp. 377–381.

36. Granda et al., "Controlling an Organic Synthesis Robot."

37. Lowe, D., "AI Designs Organic Syntheses." *Nature,* 2018. **555**(7698): pp. 592–593.

38. Simonite, T., "Machine Vision Helps Spot New Drug Treatments," *MIT Technology Review.* 2017.

39. Xiong, H.Y., et al., "The Human Splicing Code Reveals New Insights into the Genetic Determinants of Disease." *Science,* 2015. **347**(6218): p. 1254806.

40. "Atomwise Opens Applications for Historic AI Drug Discovery Awards," *Atomwise.* 2017.

41. Gershgorn, D., "Artificial Intelligence Could Build New Drugs Faster Than Any Human Team," *Quartz.* 2017.

42. Schneider, G., "Automating Drug Discovery." *Nat Rev Drug Discov,* 2018. **17**(2): pp. 97–113.

43. Kurtzman, L., "Public-Private Consortium Aims to Cut Preclinical Cancer Drug Discovery from Six Years to Just One," *UCSF News Center.* 2017.

44. Nelson, "Artificial Intelligence Could Drastically Reduce the Time."

45. Hernandez, D., "How Robots Are Making Better Drugs, Faster," *Wall Street Journal.* 2018.

46. Chakradhar, S., "Predictable Response: Finding Optimal Drugs and Doses Using Artificial Intelligence." *Nat Med,* 2017. **23**(11): pp. 1244–1247.

47. Maney, K., "AI Promises Life-Changing Alzheimer's Drug Breakthrough," *Newsweek.* 2018.

48. Comstock, J., "Benevolent AI Gets $115M to Harness AI for New Drug Discovery," *MobiHealthNews.* 2018.

49. Robie, A. A., et al., "Mapping the Neural Substrates of Behavior." *Cell,* 2017. **170**(2): pp. 393–406 e28.

50. Dasgupta, S., C. F. Stevens, and S. Navlakha, "A Neural Algorithm for a Fundamental Computing Problem." *Science,* 2017. **358**(6364): pp. 793–796.

51. Savelli, F., and J. J. Knierim, "AI Mimics Brain Codes for Navigation." *Nature,* 2018. **557**(7705): pp. 313–314; Abbott, A., "AI Recreates Activity Patterns That Brain Cells Use in Navigation," *Nature.* 2018; Beall, A., "Deep-Mind Has Trained an AI to Unlock the Mysteries of Your Brain," *Wired.* 2018; Banino, A., et al., "Vector-Based Navigation Using Grid-Like Representations in Artificial Agents." *Nature,* 2018. **557**(7705): pp. 429–433.

52. Koch, C., "To Keep Up with AI, We'll Need High-Tech Brains," *Wall Street Journal.* 2013.

53. Hassabis, D., et al., "Neuroscience-Inspired Artificial Intelligence." *Neuron,* 2017. **95**(2): pp. 245–258.

54. Cherry, K. M., and L. Qian, "Scaling Up Molecular Pattern Recognition with DNA-Based Winner-Take-All Neural Networks." *Nature,* 2018. **559**(7714): pp. 370–376.

55. Jain, V., and M. Januszewski, "Improving Connectomics by an Order of Magnitude," *Google AI Blog.* 2018; Januszewski, M., et al., "High-Precision Automated Reconstruction of Neurons with Flood-Filling Networks." *Nat Methods,* 2018. **15**(8): pp. 605–610.

56. "Japan's K Supercomputer," *Trends in Japan.* 2012.

57. Luo, L., "Why Is the Human Brain So Efficient?," *Nautil.us.* 2018.

58. "Neural Networks Are Learning What to Remember and What to Forget," *MIT Technology Review.* 2017.

59. Aljundi, R., et al., *Memory Aware Synapses: Learning What (Not) to Forget,* bioRxiv. 2017.

60. Koch, C., "To Keep Up with AI, We'll Need High-Tech Brains," *Wall Street Journal.* 2017; "Cell Types," in *Allen Brain Atlas.* 2018. Seattle, WA: Allen Institute Publications for Brain Science.

61. Waldrop, M. M., "Neuroelectronics: Smart Connections." *Nature,* 2013. **503**(7474): pp. 22–24.

62. Condliffe, J., "AI-Controlled Brain Implants Help Improve People's Memory." *MIT Technology Review.* 2018; Carey, B., "The First Step Toward a Personal Memory Maker?," *New York Times.* 2018.

63. Broccard, F. D., et al., "Neuromorphic Neural Interfaces: From Neurophysiological Inspiration to Biohybrid Coupling with Nervous Systems." *J Neural Eng,* 2017. **14**(4): p. 041002.

64. Metz, C., "Chips Off the Old Block: Computers Are Taking Design Cues from Human Brains," *New York Times.* 2017.

65. Ambrogio, S., et al., "Equivalent-Accuracy Accelerated Neural-Network Training Using Analogue Memory." *Nature,* 2018. **558**(7708): pp. 60–67; Moon, M., "'Artificial Synapse' Points the Way Toward Portable AI Devices," *Engadget.* 2018.

66. Christiansen, E., "Seeing More with In Silico Labeling of Microscopy Images," *Google AI Blog.* 2018; Grens, K., "Deep Learning Allows for Cell Analysis Without Labeling," *Scientist.* 2018; Christiansen, E. M., et al., "In Silico Labeling: Predicting Fluorescent Labels in Unlabeled Images." *Cell,* 2018. **173**(3): pp. 792–803 e19.

67. Grens, "Deep Learning Allows for Cell Analysis Without Labeling"; Sullivan, D. P., and E. Lundberg, "Seeing More: A Future of Augmented Microscopy." *Cell,* 2018. **173**(3): pp. 546–548.

68. Ounkomol, C., et al., "Label-Free Prediction of Three-Dimensional Fluorescence Images from Transmitted-Light Microscopy." *Nat Methods,* 2018.

69. Sullivan, D. P., et al., "Deep Learning Is Combined with Massive-Scale Citizen Science to Improve Large-Scale Image Classification." *Nat Biotechnol,* 2018. **36**(9): pp. 820–828.

70. Ota, S., et al., "Ghost Cytometry." *Science,* 2018. **360**(6394): pp. 1246–1251.

71. Nitta, N., et al., "Intelligent Image-Activated Cell Sorting." *Cell,* 2018. **175**(1): pp. 266–276 e13.

72. Weigert, M., et al., *Content-Aware Image Restoration: Pushing the Limits of Fluorescence Microscopy,* bioRxiv. 2017; Yang, S. J., et al., "Assessing Microscope Image Focus Quality with Deep Learning." *BMC Bioinformatics,* 2018. **19**(1): p. 77.

73. Ouyang, W., et al., "Deep Learning Massively Accelerates Super-Resolution Localization Microscopy." *Nat Biotechnol,* 2018. **36**(5): pp. 460–468.

74. Stumpe, M., "An Augmented Reality Microscope for Realtime Automated Detection of Cancer," *Google AI Blog.* 2018.

75. Wise, J., "These Robots Are Learning to Conduct Their Own Science Experiments," *Bloomberg.* 2018.

76. Bohannon, J., "A New Breed of Scientist, with Brains of Silicon," *Science Magazine.* 2017.

77. Appenzeller, "The Scientists' Apprentice."

78. Butler, K. T., et al., "Machine Learning for Molecular and Materials Science." *Nature,* 2018. **559**(7715): pp. 547–555.

CHAPTER 11: DEEP DIET

1. Estruch, R., et al., "Primary Prevention of Cardiovascular Disease with a Mediterranean Diet Supplemented with Extra-Virgin Olive Oil or Nuts." *N Engl J Med,* 2018. **378**(25): pp. e34; "Ioannidis: Most Research Is Flawed; Let's Fix It." *Medscape One-on-One,* 2018. https://www.medscape.com/viewarticle/898405.

2. Estruch et al., "Primary Prevention of Cardiovascular Disease."

3. Ioannidis, J. P. A., and J. F. Trepanowski, "Disclosures in Nutrition Research: Why It Is Different." *JAMA,* 2018. **319**(6): pp. 547–548.

4. Penders, B., "Why Public Dismissal of Nutrition Science Makes Sense: Post-Truth, Public Accountability and Dietary Credibility." *British Food Journal,* 2018. https://doi.org/10.1108/BFJ-10-2017-0558.

5. Dehghan, M., et al., "Associations of Fats and Carbohydrate Intake with Cardiovascular Disease and Mortality in 18 Countries from Five Continents (PURE): A Prospective Cohort Study." *Lancet,* 2017. **390**(10107): pp. 2050–2062.

6. Micha, R., et al., "Association Between Dietary Factors and Mortality from Heart Disease, Stroke, and Type 2 Diabetes in the United States." *JAMA,* 2017. **317**(9): pp. 912–924.

7. Bertoia, M. L., et al., "Changes in Intake of Fruits and Vegetables and Weight Change in United States Men and Women Followed for Up to 24 Years: Analysis from Three Prospective Cohort Studies." *PLoS Med,* 2015. **12**(9): p. e1001878.

8. Aune, D., et al., "Whole Grain Consumption and Risk of Cardiovascular Disease, Cancer, and All Cause and Cause Specific Mortality: Systematic Review and Dose-Response Meta-Analysis of Prospective Studies." *BMJ,* 2016. **353**: p. i2716.

9. Gunter, M. J., et al., "Coffee Drinking and Mortality in 10 European Countries: A Multinational Cohort Study." *Ann Intern Med,* 2017. **167**(4): pp. 236–247; Poole, R., et al., "Coffee Consumption and Health: Umbrella Review of Meta-Analyses of Multiple Health Outcomes." *BMJ,* 2017. **359**: p. j5024; Loftfield, E., et al., "Association of Coffee Drinking with Mortality by Genetic Variation in Caffeine Metabolism: Findings from the UK Biobank." *JAMA Intern Med,* 2018. **178**(8): pp. 1086–1097; Park, S. Y., et

al., "Is Coffee Consumption Associated with Lower Risk for Death?" *Ann Intern Med,* 2017. **167**(4). http://annals.org/aim/fullarticle/2643437/coffee -consumption-associated-lower-risk-death; Park, S. Y., et al., "Association of Coffee Consumption with Total and Cause-Specific Mortality Among Non-white Populations." *Ann Intern Med,* 2017. **167**(4): pp. 228–235.

10. Schoenfeld, J. D., and J. P. Ioannidis, "Is Everything We Eat Associ-ated with Cancer? A Systematic Cookbook Review." *Am J Clin Nutr,* 2013. **97**(1): pp. 127–134.

11. Dehghan, M., et al., "Association of Dairy Intake with Cardiovascular Disease and Mortality in 21 Countries from Five Continents (PURE): A Pro-spective Cohort Study." *Lancet,* 2018. **392**(10161): pp. 2288–2297; Mente, A., et al., "Urinary Sodium Excretion, Blood Pressure, Cardiovascular Dis-ease, and Mortality: A Community-Level Prospective Epidemiological Cohort Study." *Lancet,* 2018. **392**(10146).

12. Belluz, J., and J. Zarracina, "Sugar, Explained," *Vox.* 2017.

13. Taubes, G., "Big Sugar's Secret Ally? Nutritionists," *New York Times.* 2017.

14. McGandy, R. B., D. M. Hegsted, and F. J. Stare, "Dietary Fats, Carbohydrates and Atherosclerotic Vascular Disease." *N Engl J Med,* 1967. **277**(4): pp. 186–192.

15. Nestle, M., "Food Politics," *Food Politics.* 2017.

16. Messerli, F., "Salt and Heart Disease: A Second Round of 'Bad Sci-ence'?" *Lancet,* 2018. **392**(10146): pp. 456–458.

17. Messerli, "Salt and Heart Disease." Mente, A., et al., "Urinary So-dium Excretion, Blood Pressure, Cardiovascular Disease, and Mortality: A Community-Level Prospective Epidemiological Cohort Study." *Lancet,* 2018. **392**(10146): pp. 496–506.

18. Messerli, "Salt and Heart Disease."

19. Jones, B., "Sorry, DNA-Based Diets Don't Work," *Futurism.* 2018.

20. Gardner, C. D., et al., "Effect of Low-Fat vs Low-Carbohydrate Diet on 12-Month Weight Loss in Overweight Adults and the Association with Genotype Pattern or Insulin Secretion: The DIETFITS Randomized Clinical Trial." *JAMA,* 2018. **319**(7): pp. 667–679.

21. Chambers, C., "Mindless Eating: Is There Something Rotten Behind the Research?," *Guardian.* 2018.

22. Zeevi, D., et al., "Personalized Nutrition by Prediction of Glycemic Responses." *Cell,* 2015. **163**(5): pp. 1079–1094.

23. Segal, E., and E. Elinav, *The Personalized Diet: The Pioneering Program to Lose Weight and Prevent Disease.* 2017. New York: Grand Central Life & Style.

24. Jumpertz von Schwartzenberg, R., and P. J. Turnbaugh, "Siri, What Should I Eat?" *Cell,* 2015. **163**(5): pp. 1051–1052.

25. Korem, T., et al., "Bread Affects Clinical Parameters and Induces Gut Microbiome–Associated Personal Glycemic Responses." *Cell Metab*, 2017. **25**(6): pp. 1243–1253 e5.

26. Korem et al., "Bread Affects Clinical Parameters."

27. Segal and Elinav, *The Personalized Diet.*

28. Azad, M. B., et al., "Nonnutritive Sweeteners and Cardiometabolic Health: A Systematic Review and Meta-Analysis of Randomized Controlled Trials and Prospective Cohort Studies." *CMAJ*, 2017. **189**(28): pp. E929–E939.

29. Segal and Elinav, *The Personalized Diet.*

30. Segal and Elinav, *The Personalized Diet.*

31. Hulman, A., et al., "Glucose Patterns During an Oral Glucose Tolerance Test and Associations with Future Diabetes, Cardiovascular Disease and All-Cause Mortality Rate." *Diabetologia*, 2018. **61**(1): pp. 101–107.

32. Martin, A., and S. Devkota, "Hold the Door: Role of the Gut Barrier in Diabetes." *Cell Metab*, 2018. **27**(5): pp. 949–951; Thaiss, C. A., et al., "Hyperglycemia Drives Intestinal Barrier Dysfunction and Risk for Enteric Infection." *Science,* 2018. **359**(6382): pp. 1376–1383.

33. Wu, D., et al., "Glucose-Regulated Phosphorylation of TET2 by AMPK Reveals a Pathway Linking Diabetes to Cancer." *Nature,* 2018. **559**(7715): pp. 637–641.

34. Hall, H., et al., "Glucotypes Reveal New Patterns of Glucose Dysregulation." *PLoS Biol,* 2018. **16**(7): p. e2005143.

35. Albers, D. J., et al., "Personalized Glucose Forecasting for Type 2 Diabetes Using Data Assimilation." *PLoS Comput Biol,* 2017. **13**(4): p. e1005232; Liu, F., et al., "Fructooligosaccharide (FOS) and Galactooligosaccharide (GOS) Increase Bifidobacterium but Reduce Butyrate Producing Bacteria with Adverse Glycemic Metabolism in Healthy Young Population." *Sci Rep,* 2017. **7**(1): p. 11789.

36. Gill, S., and S. Panda, "A Smartphone App Reveals Erratic Diurnal Eating Patterns in Humans That Can Be Modulated for Health Benefits." *Cell Metab,* 2015. **22**(5): pp. 789–798.

37. Wallace, C., "Dietary Advice from the Gut," *Wall Street Journal.* 2018. p. R6.

38. Reynolds, G., "Big Data Comes to Dieting," *New York Times.* 2018; Piening, B. D., et al., "Integrative Personal Omics Profiles During Periods of Weight Gain and Loss," *Cell Syst.* 2018.

39. Wallace, "Dietary Advice from the Gut."

40. Kalantar-Zadeh, K., "A Human Pilot Trial of Ingestible Electronic Capsules Capable of Sensing Different Gases in the Gut." *Nature Electronics,* 2018. **1**: pp. 79–87.

41. Isabella, V. M., et al., "Development of a Synthetic Live Bacterial Therapeutic for the Human Metabolic Disease Phenylketonuria," *Nat Biotechnol.* 2018.

Chapter 12: The Virtual Medical Assistant

1. "Finding a Voice," *Economist.* 2017.

2. Darrow, B., "Why Smartphone Virtual Assistants Will Be Taking Over for Your Apps Soon," *Fortune.* 2016.

3. Levy, S., "Inside Amazon's Artificial Intelligence Flywheel," *Wired.* 2018.

4. Condliffe, J., "In 2016, AI Home Assistants Won Our Hearts," *MIT Technology Review.* 2016.

5. Eadicicco, L., "Google Wants to Give Your Computer a Personality," *Time.* 2017.

6. Hempel, J., "Voice Is the Next Big Platform, and Alexa Will Own It," *Wired.* 2016.

7. Terado, T., "Why Chatbots Aren't Just a Fad," *Machine Learnings.* 2017.

8. Arndt, R. Z., "The New Voice of Patient Engagement Is a Computer," *Modern Healthcare.* 2017. pp. 20–22.

9. Carr, N., "These Are Not the Robots We Were Promised," *New York Times.* 2017.

10. Anders, G., "Alexa, Understand Me," *MIT Technology Review.* 2017.

11. Domingos, P., *Pedro Domingos Interviews with Eric Topol.* September 2017.

12. Goode, L., "How Google's Eerie Robot Phone Calls Hint at AI's Future," *Wired.* 2018.

13. Foote, A., "Inside Amazon's Painstaking Pursuit to Teach Alexa French," *Wired.* 2018.

14. Kornelis, C., "AI Tools Help the Blind Tackle Everyday Tasks," *Wall Street Journal.* 2018; Bogost, I., "Alexa Is a Revelation for the Blind." *Atlantic.* 2018; Kalish, J., "Amazon's Alexa Is Life-Changing for the Blind," *Medium.* 2018.

15. Sun, Y., "Why 500 Million People in China Are Talking to This AI," *MIT Technology Review.* 2017.

16. Hutson, M., "Lip-Reading Artificial Intelligence Could Help the Deaf—or Spies," *Science Magazine.* 2018; Shillingford, B., et al., *Large-Scale Visual Speech Recognition.* arXiv, 2018.

17. Abel, A., "Orwell's 'Big Brother' Is Already in Millions of Homes: Her Name Is Alexa," *Macleans.* 2018.

18. Applin, S. A., "Amazon's Echo Look: We're Going a Long Way Back, Baby," *Medium.* 2017.

19. Vincent, J., "Fashion Startup Stops Using AI Tailor After It Fails to Size Up Customers Correctly," *Verge.* 2018.

20. Wilson, M., "A Simple Design Flaw Makes It Astoundingly Easy to Hack Siri and Alexa," *Fast Co Design*. 2017.

21. Smith, I., "Amazon Releases Echo Data in Murder Case, Dropping First Amendment Argument," *PBS NewsHour*. 2017.

22. Shaban, H., "Amazon Echo Recorded a Couple's Conversation, Then Sent Audio to Someone They Know," *LA Times*. 2018.

23. Carr, "These Are Not the Robots We Were Promised."

24. Turkle, S., "The Attack of the Friendly Robots," *Washington Post*. 2017.

25. Tsukayama, H., "When Your Kid Tries to Say 'Alexa' Before 'Mama,'" *Washington Post*. 2017; Aubrey, A., "Alexa, Are You Safe for My Kids?," *Health Shots NPR*. 2017.

26. Kastrenakes, J., "Alexa Will Come to Headphones and Smartwatches This Year," *Verge*. 2018.

27. Muoio, D., "Voice-Powered, In-Home Care Platform Wins Amazon Alexa Diabetes Competition," *MobiHealthNews*. 2017.

28. Kiistala, M., "One Man's Quest to Cure Diabetes 2," *Forbes*. 2017.

29. Stockton, N., "Veritas Genetics Scoops Up an AI Company to Sort Out Its DNA," *Wired*. 2017.

30. Stein, N., and K. Brooks, "A Fully Automated Conversational Artificial Intelligence for Weight Loss: Longitudinal Observational Study Among Overweight and Obese Adults." *JMIR,* 2017. **2**(2): e(28).

31. Ross, C., "Deal Struck to Mine Cancer Patient Database for New Treatment Insights," *Stat News*. 2017.

32. Muoio, D., "Machine Learning App Migraine Alert Warns Patients of Oncoming Episodes," *MobiHealthNews*. 2017.

33. Comstock, J., "New ResApp Data Shows ~90 Percent Accuracy When Diagnosing Range of Respiratory Conditions," *MobiHealthNews*. 2017.

34. Han, Q., et al., *A Hybrid Recommender System for Patient-Doctor Matchmaking in Primary Care*. arXiv, 2018.

35. Razzaki, S., et al., *A Comparative Study of Artificial Intelligence and Human Doctors for the Purpose of Triage and Diagnosis*. arXiv, 2018; Olson, P., "This AI Just Beat Human Doctors on a Clinical Exam," *Forbes*. 2018.

36. Foley, K. E., and Y. Zhou, "Alexa Is a Terrible Doctor," *Quartz*. 2018.

37. "The Digital Puppy That Keeps Seniors Out of Nursing Homes (Wired)," Pace University. 2017. https://www.pace.edu/news-release/wired -digital-puppy-keeps-seniors-out-nursing-homes.

38. Lagasse, J., "Aifloo Raises $6 Million for Elder-Focused Smart Wristband," *MobiHealthNews*. 2017.

39. Chen, J. H., and S. M. Asch, "Machine Learning and Prediction in Medicine—Beyond the Peak of Inflated Expectations." *N Engl J Med,* 2017. **376**(26): pp. 2507–2509.

40. Greene, J. A., and J. Loscalzo, "Putting the Patient Back Together—Social Medicine, Network Medicine, and the Limits of Reductionism." *N Engl J Med,* 2017. **377**(25): pp. 2493–2499.

41. Duncan, D. E., "Can AI Keep You Healthy?," *MIT Technology Review.* 2017; Cyranoski, D., "Jun Wang's iCarbonX Heads Consortium Using AI in Health and Wellness." *Nat Biotechnol,* 2017. **35**(2): pp. 103–105; Cyranoski, D., "Chinese Health App Arrives." *Nature,* 2017. **541**: pp. 141–142.

42. Knight, W., "An Algorithm Summarizes Lengthy Text Surprisingly Well," *MIT Technology Review.* 2017.

43. Haun, K., and E. Topol, "The Health Data Conundrum," *New York Times.* 2017; Kish, L. J., and E. J. Topol, "Unpatients—Why Patients Should Own Their Medical Data." *Nat Biotechnol,* 2015. **33**(9): pp. 921–924.

44. Heller, N., "Estonia, the Digital Republic," *New Yorker.* 2017.

45. Goldman, B., *The Power of Kindness: Why Empathy Is Essential in Everyday Life.* 2018. New York: HarperCollins, pp. 202–203.

46. Mar, A., "Modern Love. Are We Ready for Intimacy with Androids?," *Wired.* 2017.

47. Di Sturco, G., "Meet Sophia, the Robot That Looks Almost Human," *National Geographic.* 2018.

48. Sagar, M., and E. Broadbent, "Participatory Medicine: Model Based Tools for Engaging and Empowering the Individual." *Interface Focus,* 2016. **6**(2): p. 20150092.

49. Patel, M. S., K. G. Volpp, and D. A. Asch, "Nudge Units to Improve the Delivery of Health Care." *N Engl J Med,* 2018. **378**(3): pp. 214–216.

50. Emanuel, E. J., "The Hype of Virtual Medicine," *Wall Street Journal.* 2017; Lopatto, E., "End of Watch: What Happens When You Try to Change Behavior Without Behavioral Science?," *Verge.* 2018.

51. Marteau, T. M., "Changing Minds About Changing Behaviour." *Lancet,* 2018. **391**(10116): pp. 116–117.

52. Subrahmanian, V. S., and S. Kumar, "Predicting Human Behavior: The Next Frontiers." *Science,* 2017. **355**(6324): p. 489.

53. "Individual Access to Genomic Disease Risk Factors Has a Beneficial Impact on Lifestyles," *EurekAlert!.* 2018.

54. Marteau, T. M., "Changing Minds About Changing Behaviour." *Lancet,* 2018. **391**(10116): pp. 116–117.

CHAPTER 13: DEEP EMPATHY

1. Mueller, M. S., and R. M. Gibson, *National Health Expenditures, Fiscal Year 1975.* Bulletin 1976. https://www.ssa.gov/policy/docs/ssb/v39n2/v39n2p3.pdf.

2. "Largest Private Equity and Venture Capital Health System Investors," *Modern Healthcare.* 2018.

3. Peabody, F. W., "The Care of the Patient." *MS/JAMA,* 1927. **88**: pp. 877–882.

4. Belluz, J., "Doctors Have Alarmingly High Rates of Depression. One Reason: Medical School," *Vox.* 2016; Oaklander, M., "Doctors on Life Support," *Time.* 2015; Wright, A. A., and I. T. Katz, "Beyond Burnout—Redesigning Care to Restore Meaning and Sanity for Physicians." *N Engl J Med,* 2018. **378**(4): pp. 309–311.

5. Farmer, B., "Doctors Reckon with High Rate of Suicide in Their Ranks," *Kaiser Health News.* 2018.

6. Andreyeva, E., G. David, and H. Song, *The Effects of Home Health Visit Length on Hospital Readmission.* 2018, National Bureau of Economic Research.

7. Maldonado, M., "Is This How It's Supposed to Be?" *Ann Intern Med,* 2018. **169**(5): pp. 347–348.

8. Tingley, K., "Trying to Put a Value on the Doctor-Patient Relationship," *New York Times.* 2018.

9. Linzer, M., et al., "Joy in Medical Practice: Clinician Satisfaction in the Healthy Work Place Trial." *Health Aff* (Millwood), 2017. **36**(10): pp. 1808–1814.

10. Whillans, A. V., et al., "Buying Time Promotes Happiness." *Proc Natl Acad Sci U S A,* 2017. **114**(32): pp. 8523–8527.

11. Schulte, B., "Time in the Bank: A Stanford Plan to Save Doctors from Burnout," *Washington Post.* 2015.

12. Rosenthal, D. I., and A. Verghese, "Meaning and the Nature of Physicians' Work." *N Engl J Med,* 2016. **375**(19): pp. 1813–1815.

13. Darzi, A., H. Quilter-Pinner, and T. Kibasi, "Better Health and Care for All: A 10-Point Plan for the 2020s. The Final Report of the Lord Darzi Review of Health and Care," *IPPR.* 2018.

14. Wright and Katz, "Beyond Burnout."

15. Epstein, R. M., and M. R. Privitera, "Doing Something About Physician Burnout." *Lancet,* 2016. **388**(10057): pp. 2216–2217.

16. Tahir, D., "Doctors Barred from Discussing Safety Glitches in U.S.-Funded Software," *Politico.* 2015.

17. Madara, J. L., and D. M. Hagerty, *AMA 2017 Annual Report. Collaboration. Innovation. Results.* 2018, American Medical Association.

18. Ballhaus, R., "Michael Cohen's D.C. Consulting Career: Scattershot, with Mixed Success," *Wall Street Journal.* 2018.

19. Castle, M., "Matthew Castle: Burnout," *BMJ Opinion.* 2017.

20. el Kaliouby, R., "We Need Computers with Empathy," *MIT Technology Review.* 2017.

21. Mar, A., "Modern Love: Are We Ready for Intimacy with Androids?," *Wired.* 2017.

22. Derksen, F., J. Bensing, and A. Lagro-Janssen, "Effectiveness of Empathy in General Practice: A Systematic Review." *Br J Gen Pract,* 2013. **63**(606): pp. e76–e84.

23. Rosenthal and Verghese, "Meaning and the Nature of Physicians' Work."

24. Kelm, Z., et al., "Interventions to Cultivate Physician Empathy: A Systematic Review." *BMC Med Educ,* 2014. **14**: p. 219.

25. Scales, D., "Doctors Have Become Less Empathetic, but Is It Their Fault?," *Aeon Ideas.* 2016.

26. Denworth, L., "I Feel Your Pain," *Scientific American.* 2017.

27. Valk, S. L., et al., "Structural Plasticity of the Social Brain: Differential Change After Socio-Affective and Cognitive Mental Training." *Sci Adv,* 2017. **3**(10): p. e1700489.

28. "Presence: The Art & Science of Human Connection." Stanford Medicine. August 14, 2018. http://med.stanford.edu/presence.html.

29. Verghese, A., "The Importance of Being." *Health Aff* (Millwood), 2016. **35**(10): pp. 1924–1927.

30. Roman, S., "Sharon Roman: In Good Hands," *BMJ Opinion.* 2017.

31. Mauksch, L. B., "Questioning a Taboo: Physicians' Interruptions During Interactions with Patients." *JAMA,* 2017. **317**(10): pp. 1021–1022.

32. Manteuffel, R., "Andrea Mitchell Remembers What It Was Like Being Carried Out of a News Conference," *Washington Post.* 2018.

33. Kneebone, R., "In Practice: The Art of Conversation." *Lancet,* 2018.

34. Corcoran, K., "The Art of Medicine: Not Much to Say Really." *Lancet,* 2018. **391**(10133).

35. Schoen, J., "The Incredible Heart of Mr. B." *Ann Intern Med,* 2017. **166**(6): pp. 447–448.

36. McCarron, T. L., M. S. Sheikh, and F. Clement, "The Unrecognized Challenges of the Patient-Physician Relationship." *JAMA Intern Med,* 2017. **177**(11): pp. 1566–1567.

37. Iglehart, J. K., "'Narrative Matters': Binding Health Policy and Personal Experience." *Health Affairs,* 1999. **18**(4). https://www.healthaffairs.org/doi/10.1377/hlthaff.18.4.6.

38. Schoen, "The Incredible Heart of Mr. B"; Molitor, J. A., "A Great Gift." *Ann Intern Med,* 2017. **167**(6): p. 444; Al-Shamsi, M., "Moral Dilemma in the ER." *Ann Intern Med,* 2017. **166**(12): pp. 909–910; Goshua, G., "Shared Humanity." *Ann Intern Med,* 2017. **167**(5): p. 359.

39. Rowland, K., "You Don't Know Me." *Lancet,* 2017. **390**: pp. 2869–2870.

40. Awdish, R. L. A., and L. L. Berry, "Making Time to Really Listen to Your Patients," *Harvard Business Review.* 2017.

41. Wheeling, K., "How Looking at Paintings Became a Required Course in Medical School," *Yale Medicine.* 2014.

42. Verghese, "The Importance of Being."

43. Gurwin, J., et al., "A Randomized Controlled Study of Art Observation Training to Improve Medical Student Ophthalmology Skills." *Ophthalmology,* 2018. **125**(1): pp. 8–14.

44. Epstein, D., and M. Gladwell, "The Temin Effect." *Ophthalmology,* 2018. **125**(1): pp. 2–3.

45. Parker, S., "Two Doctors Meet." *Ann Intern Med,* 2018. **168**(2): p. 160.

46. Jurgensen, J., "A Show Redefines the TV Hero," *Wall Street Journal.* 2017.

47. Verghese, A., "Treat the Patient, Not the CT Scan," *New York Times.* 2011.

48. Wiebe, C., "Abraham Verghese: 'Revolution' Starts at Bedside," *Medscape.* 2017.

49. Verghese, A., "A Touch of Sense," *Health Affairs.* 2009.

50. Aminoff, M. J., "The Future of the Neurologic Examination." *JAMA Neurol,* 2017. **74**(11): pp. 1291–1292.

51. Hall, M. A., et al., "Trust in Physicians and Medical Institutions: What Is It, Can It Be Measured, and Does It Matter?" *Milbank Q,* 2001. **79**(4): pp. 613–639. https://www.ncbi.nlm.nih.gov/pubmed/11789119.

52. Reddy, S., "How Doctors Deliver Bad News," *Wall Street Journal.* 2015.

53. Ofri, D., "The Art of Medicine: Losing a Patient." *Lancet,* 2017. **389**: pp. 1390–1391.

54. "The Pharos of Alpha Omega Alpha Honor Medical Society." *Pharos,* 2016. **79**(1): pp. 1–64.

55. Kaplan, L. I., "The Greatest Gift: How a Patient's Death Taught Me to Be a Physician." *JAMA,* 2017. **318**(18): pp. 1761–1762.

56. Verghese, A., "The Way We Live Now: 12-8-02; The Healing Paradox," *New York Times Magazine.* 2002.

57. Tingley, "Trying to Put a Value on the Doctor-Patient Relationship."

58. "2017 Applicant and Matriculant Data Tables," *Association of American Medical Colleges.* 2017.

59. Freeman, S., et al., "Active Learning Increases Student Performance in Science, Engineering, and Mathematics." *Proc Natl Acad Sci USA,* 2014. **111**(23): pp. 8410–8415.

60. Awdish, R. L. A., "The Critical Window of Medical School: Learning to See People Before the Disease," *NEJM Catalyst.* 2017.

61. Stock, J., "Does More Achievement Make Us Better Physicians? The Academic Arms Race." *JAMA Intern Med,* 2018. **178**(5): pp. 597–598.

62. Topol, E., *The Patient Will See You Now.* 2015. New York: Basic Books.

63. Warraich, H. J., "For Doctors, Age May Be More Than a Number," *New York Times.* 2018.

INDEX

ERIC TOPOL is a world-renowned cardiologist, executive vice president of Scripps Research, founder of a new medical school, and one of the top ten most-cited medical researchers. The author of *The Patient Will See You Now* and *The Creative Destruction of Medicine*, he lives in La Jolla, California.